Dynamic Farmers' Marketing

DATE DUE	

DEMCO, INC. 38-2931

Dynamic Farmers' Marketing

A Guide to Successfully Selling
Your
Farmers' Market Products

Jeff Ishee

Bittersweet Farmstead, P.O. Box 52, Middlebrook, Virginia 24459

Dynamic Farmers' Marketing:
A Guide to Successfully Selling
***Your* Farmers' Market Products**

Copyright © 1997
by Jeffrey Wayne Ishee

Second printing
August, 2001

All rights reserved.
Printed in the USA.
First Edition
ISBN   0-9656890-0-X
Library of Congress Catalog Card Number: 96-95299

See last page for ordering information.

Warning - Disclaimer

Table of Contents

To all of the small-scale farmers,
market gardeners, and farmers' market vendors
who purchase this book.

Let's don't make any bones about it.
You have helped this farmer make a living simply
by purchasing this book.

I thank you . . . and dedicate this book to *your success at the market*.

Acknowledgments

I have not attempted to cite in the text of *Dynamic Farmers' Marketing* all of the authorities and sources consulted in the preparation of this book. To do so would require an immense amount of space. Dozens of people have assisted me along the way, and I would like to acknowledge just a few.

Special thanks go to Betty Jo Hamilton and Joel Salatin. Both are first-rate farmers, both are first-rate writers, and, luckily, both are neighbors and genuine friends of mine. Thanks for being a real inspiration to me.

I want to credit all of the farmers at the Staunton/Augusta Farmers' Market in Staunton, Virginia. Your incredible success and dynamic methods of marketing led to the concept of this book. To tell the truth, this is *your* book.

Mr. Homer Quann, a legend in farm radio broadcasting for almost fifty years, has been more than an inspiration. Homer took me under his wing and has taught me volumes about agribusiness, telling the stories of farmers, and about dealing with people. Now, via WSVA in Harrisonburg, Virginia, Homer and I produce the only daily, local, farm radio show in the mid-Atlantic. It is a blessing to work with a legend, someone who is passionate about agriculture and the people who are involved in it.

I must also thank my family for the support they have given me during this project. My dad, J. Q. Ishee, grew up on a farm in Mississippi in the height of the depression early in this century. He recently told me "Jeff, when I was young, the ultimate goal in farming was to work your way *off* the farm. I'm so glad that things have changed." Amen.

Without my wife Sheila, and children Jennifer, Matt, and Caleb, this effort would have been a total loss. Many a day they picked up where I left off, and gave me the time to work on the book. Thanks for your patience, your love, and most of all, *thanks for feeding the chickens*!

Thanks to these people, and thanks to many others for making this book possible.

Foreword

by Joel Salatin

My love affair with farming and direct marketing began when 50 multi-colored balls of fluff arrived in the mail when I was 10 years old. These chicks, the classic "assorted, as hatched, heavy breed special," catapulted me into the small farmer's dilemma: too few to sell in regular markets but too many for the family to eat.

Thus began weekly egg deliveries on my bicycle to the neighbors who had terminated their backyard flocks for the sake of "efficiency." The little chicken houses made good storage buildings. But this change in farming focus created a niche opportunity for a 10-year-old entrepreneur.

With low overhead and selling retail, the enterprise made a profit and expanded in true capitalistic style until within a couple of years I was producing 10 dozen a day. At that time, Staunton's "Curb Market" was still functioning. It was started during the depression to give cash-starved farmers an outlet for their wares, and at one time boasted nearly 50 vendors and, according to old-timers, was the busiest place in town on Saturday mornings.

By about 1970 it had dwindled to two elderly matrons who faithfully brought their wares every Saturday morning, year round. The market opened at 6:00 a.m. and went until 11:00 a.m. Dad, whose love affair with economics and accounting made him focus on the farm as a business, realized the potential -- indeed the necessity -- of receiving the retail prices in order to be economically viable.

We decided the Staunton Curb Market provided the best venue for our direct marketing. The market enjoyed special exemptions from health department and inspection requirements because it operated under the authority of the extension service. Vendors were required to be members in a "Home Demonstration Club," or what are today "Extension Homemakers' Clubs." That seemed an odd requirement for a boy so after some deliberation, the powers decided that if I joined 4-H that would be equivalent.

From 14 - 18 years old, then, I woke early every Saturday morning to man the stand at the market. What an invaluable experience! I realize now what a rare privilege it was for me to participate in such an unrestricted marketing opportunity as late as the mid-70's. We sold eggs, fresh beef and pork, my brother's fresh rabbit, and extra dairy products from our family Guernsey milk cows -- butter, buttermilk, cottage cheese and yogurt. I added a special garden and sold vegetables during the summer.

When I went off to college in 1975, we closed the stand and shortly thereafter the two ladies stopped as well, closing down the market and all the special "grandfather" privileges we enjoyed. I've always thought that perhaps this enthusiastic youngster squeezed a little more time out of these two grandmothers. One sold baked items and made potato salad with sweet pickles in it -- guess what I always had for breakfast? If she didn't bring potato salad, a hefty wedge of her pound cake and a thermos of raw Guernsey milk were just about as close to paradise as you could get. The other lady came from a diversified farmstead, and sold cured pork, a few eggs, baked items and loads of vegetables.

I often wonder what would have happened had I not gone to college, but stayed at the market to continue the privileges we enjoyed. Looking back, I realize we were at the brink of a new time. Rachel Carson's *Silent Spring* had only been out for a decade and the environmental movement, fueled by the anti-conventional sentiment of the Vietnam era, had not yet developed into mainstream buying power. Just 5 years later, after returning to the farm, I was amazed at the change in public awareness concerning humane animal husbandry, clean food and ecological farming practices. I cranked up our farm's direct marketing angle again, using many of the skills I learned as a kid. Today, we enjoy a full-time farming enterprise that markets salad bar beef, pastured poultry, pastured eggs, pigaerator pork, pastured rabbit, pastured turkeys, vegetables and forestry products to about 400 patrons.

The new climate has now spawned the farmers' market movement, fueled by both customers demanding higher quality food than can be produced on far away industrial mega-farms, and by a new group of farmers and homesteaders who enjoy people and appreciate the need to receive retail prices in order to stay economically viable.

This resurgence nationwide requires new production and marketing paradigms. Farmers for too long have been told: "You just produce it and we'll take care of the rest." Following that philosophy, farmers now receive only 9 cents of the supermarket dollar. Many farmers understand that we must get more than 9 cents. But the skills, techniques and philosophical base to take back that market share are woefully lacking in the farming community.

Whether we have a backyard garden of flowers and vegetables, or a multi-acre livestock farm, farmers' markets bring to us unprecedented opportunities to build food-farm relationships and access retail dollars. We have now rejoined our town's farmers' market, and although it is only seasonal and does not enjoy the freedoms it once did, its magnetism is still irresistible.

A whole parking lot full of vendors and hundreds of customers attest to a cultural change that is thrilling to behold. The immeasurable dynamic energy of a farmers' market radiates into the country and the city. Jeff's book captures the essence and energy from the farmers' market experience, and helps to channel its creativity. Whether you are a concerned consumer wanting more food choice, or a small farmer desiring a better marketing venue, or anyone in between, this book will stimulate your thinking.

We all need to be energized in our collective effort to produce and market better food to informed and deserving patrons. I hope you will let Jeff's book be a catalyst in your own thinking as you dream, as you develop a vision of your own opportunity in the food-farm connection. Let's get to it.

Joel Salatin
March 1997

Introduction

This book was born out of my fascination with two subjects: farming and marketing. Since several of my free lance articles on both topics were published in local and regional markets, I was inspired to venture further and put together this work. My sister Beverly has always said "Jeff, you should write a book!" (I think she envisioned a novel of some sort. Tom Clancy I'm not.) My wife Sheila prodded me on, and fellow vendors at the Staunton/Augusta Farmers' Market in Staunton, Virginia encouraged me also.

The single thing that I hope makes this book different from most others on the subject is the fact it was written, published, marketed, and distributed from a small farm. This certainly does not make it an agricultural commodity, but I can tell you that it certainly is *one* of our own *farm products*.

Writing in the fall of the year after my work in the market gardens has slowed dramatically allows me to concentrate on the project at hand. Nevertheless, as most of my readers will be acutely aware, work never completely ceases on a farmstead. At the end of this paragraph, for instance, I must take a short break and go outside to collect eggs and care for the chickens and geese. I sincerely hope that these breaks in *my* writing don't result in breaks in the continuity of *your* reading and thought processes.

- - -

Well, I'm back indoors now. The wood stove is fired up, the coffee pot is on and the animals are content. My typing fingers are nice and limber, so let's get on with it!

Middlebrook, Virginia
November 1996

Chapter 1

Personal Satisfaction

A *half hour before the bright July sun would rise over the picturesque buildings of historic downtown, a scene occurred which has been repeated in* almost every community in this country for decades. A battered old farm truck glided into a slot at the public farmer's market. Silence returned as the vehicle was shut off after the long drive into town. The farmer and his wife

The farmers' marketing *experience* can be one of joy and satisfaction, seen clearly here on the faces of Mr. and Mrs. Rufus L. Jones at the Montgomery Curb Market in Montgomery, Alabama.

stepped out, stretched, and looked around. They spoke to their neighboring vendors at the market with a hushed "good morning," as if not to disturb the city folk, most of whom were still sleeping in their nearby apartments.

The routine was automatic. He let the tailgate down and started sorting out the baskets and crates. She set up the old, yet reliable, folding tables along the sidewalk and covered them with worn, blue gingham tablecloths.

Across the parking lot, there were some residents of the city who had risen at an exceedingly early hour, and they had a purpose for doing so. They sat in their darkened cars as if eagerly anticipating a signal. They watched as the farmer brought out two bushels of peas, three of green beans, one of plump green bell peppers, two pecks of yellow squash, and a flat of carrots that were almost fluorescent orange, even in the dim, predawn light. Then came the sweet corn. Ladies in their cars peered over the steering wheel as they watched bag after bag of Silver Queen corn appear out of the back of the farmer's pickup.

The farmer's wife then began unloading her wares. Trays of fresh baked goods were followed by dozens of colorful jars. The bread went on the middle shelf of their market stand, while the canned goods went on the top shelf. She did this so that the morning sunlight would penetrate the jelly jars giving an effect of stained glass. Ohhhh, it worked so well on bright summer mornings such as this one! Then came bags of dinner rolls, followed by apple pies, chocolate cakes, and her renowned "farmhouse biscuits." Some items were still warm from her late night efforts. The customers in their cars kept looking at their watches.

At precisely 7:00 A.M., the old clocktower bell atop a nearby downtown building announced the hour, and another farmer's market day began. Almost in perfect unison, car doors opened and customers stepped out of their vehicles with eagerness. Several people opened their trunk and retrieved shopping bags and baskets for the excursion. As customers briskly walked into the market, the farmer seemed to be caught off guard momentarily, but then strapped on his change apron, adjusted his hat, and smiled. His wife, still placing loaves of fresh bread on display, greeted their first patron of the day by saying "Good morning Mrs. Mabry. How was your week?"

Pleasant Experience

I don't know about you, friends, but this scene is one of fundamental personal satisfaction for both parties. It is the beginning of not only a sales transaction, but the continuation of a friendship. The spectacle you have just witnessed is one of increasing occurrence all across America. It's not just an anonymous consumer and anonymous clerk going about a boring routine. No, no, no! It's an *event* that both parties look forward to all week long as they await the next farmers' market.

Satisfaction

In any type of business venture, whether it is operating a pet supply store, a barber shop, a wood stove dealership, or a stall at the local farmers' market, our ultimate goal is not only financial success, but personal satisfaction. Many people in our society feel stuck in

---

"How thrilling it is to till the earth in the new spring, plant the seeds and watch them grow as if by magic. Then harvesting the bounty of my labor and sharing it with responsive and appreciative customers who give me continual feedback of approval. Having their friendship is the final satisfaction."
Andy Lee
Backyard Market Gardening

---

humdrum jobs that they wish they could leave . . . as a matter or fact, some say they would do *anything* to get out of their current employment.

Farmers' marketing is one venture I've found that offers not only personal satisfaction, but also:

- ▸ A good source of income
- ▸ A healthy working environment
- ▸ A wonderful way to interact with other people

It is an honorable occupation.

Dale Carnegie once said "Everybody in the world is seeking happiness." The obvious question is "*How* can selling tomatoes and lettuce offer *happiness*?" Well, imagine the following transaction:

Retired school teacher Helen Barfield had been told by her physician to eat more fresh vegetables and fruit, and fewer processed foods from the supermarket. "This change in your diet will help you lose weight, supply you with more vitamins and energy, and generally make you feel better" proclaimed the doctor. So Helen took the doctor's advice and visited the local farmers' market the next weekend. It was the first time she had been to the farmers' market since her husband Raleigh had passed on 4 years ago. When they were much younger, Helen and Raleigh had gone to the market almost every week.

While strolling unaccompanied along the market commons, suddenly she was hugged from behind and heard "Why Mrs. Barfield, I haven't seen you for years!" Helen turned and saw a girl she had taught in her third grade English class. But she wasn't a little girl anymore. She was an adult woman with a small child by her side.

"Linda Baker! How are you? My! You're a mature woman now, and I see you've got a little girl of your own!" said Helen.

"Yes indeed, but it's not Linda Baker anymore. Do you remember that class clown in fourth grade named Steve Simmons? Well, we were married after college, so I'm Linda

Simmons now. Next month is our tenth anniversary!" she proudly announced.

"Tenth anniversary? My! How time has flown!" Helen exclaimed as she smiled broadly and noticed that Linda and her preschool daughter were the picture of health.

Just at that moment, the little girl lunged away from her mother's side towards a farmer's stall. She announced "Look Mommy! Mr. Johnson has fresh spinach this week. Let's get some for dinner tonight!"

Fred Johnson stood casually behind his farmers' market stand observing the conversation with more than just a passing interest. Helen looked up, and said "Freddie Johnson. You're a farmer now?"

Whereupon Linda interrupted by asking "Do you know Mr. Johnson?"

"Oh yes. I taught Freddie in my English class also, but he came through four or five years before you." said Helen.

Fred Johnson acknowledged that, yes, he too had been one of Mrs. Barfield's students, and answered her question by saying "Yes indeed! I'm a farmer now. After a few years in the Army, I came home and became entranced with gardening. The gardens got bigger and bigger every year, and I had so much fun with it that I decided to pursue it part time as a market gardener. If things keep going the way they are now, I hope to be able to farm full-time within a couple of years. Our success here at the farmers' market has been great."

Linda Simmons then said "Well I suppose so! We buy fresh vegetables and eggs from you every week Mr. Johnson, but I didn't know that we went to the same school . . . and had the same English teacher!"

The chit chat continued for some time, and then Linda Simmons bought her usual weekly items from Fred Johnson, which included: tomatoes, carrots, Irish potatoes, fresh brown eggs, and, of course, spinach. Helen Barfield also bought a few things including Romaine lettuce, yellow squash, green beans, and some snow peas. She was sure that her family doctor would be proud, because it was obvious that Fred Johnson's produce was fresh and of high quality. Never once did either party ask about price. Everyone was happy.

Relationships Developed

Scenes like this occur frequently at farmers' markets all over our nation. Rather than an anonymous trip to the newest million dollar "superstore" in town, this trip to the local farmers' market had renewed an old relationship, supported local agriculture, and provided fresh and healthy food for a grateful, new customer.

What makes *you* happy may be different, however. Abe Lincoln once observed that "most folks are about as happy as they want to be." I've seen some people working third shift on the assembly lines for minimum wage and they were as happy as a lark. And I've also seen some people in the same situation that acted like robots . . . totally without pleasure

4

or satisfaction.

Farmers' marketing can be very tiring at times, as you know if you've ever done it. But the personal satisfaction that can be gained from the experience is . . . well, it's simply unbeatable.

It is immensely satisfying to be a farmers' marketer. You are able to deal with people on friendly terms and develop those special relationships so necessary to success . . . and *they want* what *you have*!

"It takes some time to build relationships. We've had many customers out (to our farm) for dinner. It makes you vulnerable to folks. But the rewards for ministering truth, information, mutual respect and appreciation - for building bridges - are worth every bit of the effort. The bridge of course is one that goes both ways, and when they need us as much as we need them, the joy of farming could not be sweeter."

Joel Salatin
Salad Bar Beef

PERSONAL SATISFACTION
via
DYNAMIC FARMERS' MARKETING

- ► Resolve the following issues:
  - What do you want to get out of your farmers' market business?
  - What do you hope to be doing five years from now?
  - What size business do you want to operate?
  - What difficult personal decisions have you been putting off?
- ► Write down your personal goals and review them frequently.
- ► Remember what is important to you. Then take small steps towards that end.
- ► If it will not ultimately make you happier, then don't do it.

**If you have joy in your heart, your customers will know it.
Mr. Nelson Koiner assists a customer at the Harrisonburg
Farmers' Market in the Shenandoah Valley of Virginia.**

Chapter 2

Increasing Demand For Local Farm Products

etween 1994 and 1996, the USDA reported more than 700 *new* farmers' markets nationwide - from 1700 markets to more than 2400. Now does this include every Tom, Dick, and Harry who decides to call his little private roadside stand a "farmers' market"?

No, according to an extensive survey completed in July 1996. The term "farmers' market" was defined as:

"A common facility or area where several farmers/growers gather on a regular, recurring basis to sell a variety of fresh fruits and vegetables and other farm products directly to consumers."

I like it.

Without a market, you cannot have a farm. Without a farm, you cannot have a market.

Big Demand

It has been estimated that fruit and vegetable sales through direct marketing (not just farmers' markets, but all methods of direct marketing) exceed $1.1 billion, which of course provides essential income to producers like you and I! Let's think about that for a minute. One point one billion dollars . . . not million, but *billion*. Just how much is $1,000,000,000?

Well, I've often thought of it this way - the largest denomination of tender that I've ever seen was a thousand-dollar bill. And, you'd better believe that it got my attention! There,

SIGNS LIKE THIS ARE SEEN MORE AND MORE FREQUENTLY ALL ACROSS THE U.S.A. THIS SIGN, UNDER A CANOPY OF OLD OAK TREES, IS AT THE ENTRANCE TO THE MARKET IN MOBILE, ALABAMA.

in one little piece of paper, represented one **thousand** dollars, which is a rather significant amount to most family farmers around the country. But if I had a *suitcase* full of thousand-dollar bills, if I could just get my hands on *one thousand* thousand-dollar bills, well, then I'd have a million bucks! But conceivably, what if there were *a thousand* suitcases, all full of thousand-dollar bills! A stack of a thousand suitcases would make a pile as big as most suburban homes . . . and it would be <u>solid cash</u> . . . solid *thousand dollar bills*! It's really just too much money for one farmer to think about. So, let's break it down.

In 1996, there were 2410 farmers' markets listed by the USDA as being active. That

equals an average of 48 markets per state. The national average is 30 farmers per market. So let's do some math: $1.1 billion divided by 2410 = $456,431. This amount divided by 30 = $15,214.00 per farmer. Do you see the potential here?

But all of this is based on USDA estimates, not actual numbers. And, also, this estimate is for *all* methods of direct marketing, not just farmers' markets.

So what about farmers' markets? Indeed, many loosely organized markets do not even keep records of their farmers' sales. They either collect no fee at all, or collect a fee based on round figures. (See appendix D for

8

sample)

But the majority of markets *do* keep accurate records. In an actual 1994 survey (not an estimate) of 382 farmers' markets, (serving a total of 11,283 farmers) the markets reported a combined annual sales figure of $70,286,636. This results in an average yearly gross per farmer of $6229. For most of us that market on a part-time basis, this is certainly a more realistic figure than that estimated previously.

Filling A Void

Seventy-three percent of customers surveyed indicated that they drive less then ten miles to get to a good farmers' market. But the same statistic shows us that one of every four customers is so dedicated that they *would* drive in *excess* of ten miles to get to a market. And, also found in the survey is the eye-opening reality that 5 percent would even drive up to *fifty* miles to get to the market!

Without a doubt, farmers' markets are filling a void. They provide a unique experience for both farmers and customers.

In the largest cities, farmers' markets provide access to fresh fruits and vegetables due to the scarcity of large supermarkets. All customers, whether they be from the inner city or from extreme rural America, are able to shop in a pleasant and friendly environment. They also get the chance to meet the people that grow their food! The documented growth of farmers' markets nationwide over the last ten years appears to indicate that this symbiotic relationship fills a marketing void.

Without a market, you cannot have a farm. Without a farm, you cannot have a market. This simple, but undeniable, truth has been evident for centuries.

Generations ago, families produced their own provisions, and there are still some homesteaders that generate the majority of their own food. But the mass population today expects, indeed relies, on someone else to produce their food while they go about making a living. There is nothing wrong with this. It is the evolutionary path that mankind has chosen (consciously or unconsciously) to take. As agricultural technology has expanded in this century, the ratio of food producers to consumers has changed dramatically. Whereas 92% of the population in the early thirties was farm based, currently only about 2% of Americans live and work on a functional farm.

As it became possible for very few farmers to produce the bulk of the nation's food, the direct link between producer and consumer began to disappear. Filling the void were wholesalers, jobbers, and grocers. As new generations moved further and further away from the soil, people began to unwittingly think that Krogers really grew the lettuce, and that sausage was really produced by the IGA. They totally lost contact with the real *source* of their food.

PERHAPS NO MARKET IN THE NATION IS MORE CARNIVAL-LIKE THAN THE FARMERS' MARKET IN THE FRENCH QUARTER OF NEW ORLEANS.

Why People Go To Farmers' Markets

Why, indeed, *do* people shop at the local farmers' market? Supermarket chains across the nation usually offer more shopping convenience, competitive prices, exotic and imported foods, and an immensely wider selection of overall food choices. When you can have all this, and are able to shop 24 hours a day, *and* charge it to your credit card, there is no wonder that supermarkets are so popular in our modern society.

So why *is* the farmers' market industry growing so fast in recent years? Let me offer a few reasons:

▸ **Farmers' markets are like a carnival**, or even a theater, continually offering a person something new and interesting to observe and interact with. You can *always* expect more energy at a farmers' market than a supermarket. Many term their own market a "social occasion."

▸ **The market is pleasurable!** Most farmers' markets are classified as: *seasonal/fresh-air*. Take a look for a moment at those two words. Do you think there is a difference between shopping at Safeway in the month of May and shopping at the local farmers' market in May? The surge of springtime is a natural combination with farmers' market activity. The fact that most markets are fresh-air (and not a sterile, fluorescent, air-conditioned, glossy vinyl floor environment) only makes it more attractive to the typical consumer.

▸ **The market is not anonymous!** It is usually filled with people you recognize, friends and neighbors. Customers get to know their farmers on a first name basis.

▸ **Produce is usually locally grown and very fresh!** There is a measurable segment of society that not only wants fresh produce, but *demands* it. And for people with chemical sensitivities, or those that prefer organically grown fruit and vegetables, the local farmers' market may be the only source available to them.

Why Farmers Go To Farmers' Markets

Most of the conventional farmers in this country do *not* attend local farmers' markets . . . and there are many reasons why.

In the latter half of this century, emphasis in agriculture has focused not only on feeding our own towns and communities, but on feeding the world! And if I were the CEO of a corporate farm, I really couldn't find fault in this logic. After all, look at the size of the market! I wouldn't be concerned about how many bushels of green beans were going to be picked this week. I would be more concerned about the number of ships that I was going to fill with grain and send to China!

But more and more often, farmers are taking a second look at the *local needs* of agriculture, where there are less middlemen to siphon off all the profits in farming. Other reasons farmers are going back to the market include:

▸ **Farmers typically record gross sales of $200 to $600 per day** at an established market in season. And some of the bigger growers can realize more than $1000 per day!

▸ **Farmers' markets usually offer a prime location** that costs much less than a private retail outlet. Rather than taking the full burden of insurance, advertising, physical facilities, and other marketing costs, a farmer can share these expenses with others.

▸ **Customer feedback is often immediate**, offering the farmer first hand knowledge of what the shopper *really wants*. If a farmer wants to experiment with different varieties, he can try it out at the market before committing to a larger scale of production.

▸ **Minimal startup costs and immediate access for new farmers**. All you really need is a table, some baskets, and a truck to get you to the market.

- **Good advice from fellow marketers.** Farmer-to-farmer relationships are usually very positive at the market. Growers share information and successful methods they use in their own operation. And, if you need to go to the restroom or get another cup of coffee, a neighboring farmer will usually watch your stand while you are away for a few moments.

- **It's just plain fun to sell at the farmers' market.** You can joke around and have a good time. I've seen many farmers with pick-your-own operations or roadside stands who just got tired of the long arduous hours. They would comment "I always have to be here at the stand, six days a week, from 7:00 A.M. until dark. I just can't get anything done *on the farm*." But because most farmers' markets are either one or two days a week, the farmer is free not only to market his crops, but to produce them with an emphasis on quality. Most farmers look forward to "market day" because it offers them a chance to stop and talk with people, tell a story or two, and make some new friends who appreciate what they do.

- **Publicity for the farm.** Because most farmers don't utilize farmers' markets as their *sole outlet*, they use the market as publicity to gain new customers who will buy their products in other ways. These might include community supported agriculture (commonly referred to as CSA), on-farm sales, restaurants, etc.

As our nation becomes more and more conscious of the vital link between farmer and consumer, the growth of the local farmers' market movement will continue into the foreseeable future. All of the trends and indicators reveal an ever-increasing demand for fresh and healthy food. More and more, consumers want food that was harvested yesterday by a local farmer, not three weeks ago and trucked along fifteen hundred miles of interstate highway.

Chapter 3

Effective Planning and Income Potential

Have you ever thought seriously about the reproduction of an ear of corn? About the promise found in one *single kernel*? Let's take a moment and do that.

Before you can acquire that single kernel, you have to choose from a multitude of varieties in the seed catalog, or perhaps at the farm store. Included in these varieties are feed corn, silage corn, oil corn, hybrid corn, non-hybrid corn, Bt. corn, genetically altered corn, open pollinated corn, yellow corn, white corn, blue corn, Indian corn, bi-color corn, supersweet corn, sh2 corn, flour corn, and last, but not least, sweet corn. You know that you want to grow *corn*, but oh boy, does it get complex after that!

You also know that you want to sell your farm products. After all, if you don't sell what you grow or raise on the farm, then it wouldn't be a farm after all. It would be a homestead. Keeping it all for personal consumption is fine if you can homestead and focus on sustainability. I endorse the endeavor of homesteading. Nevertheless, when you sell any of your crops or livestock with the intent of gaining a profit, it ceases to be a *home*stead and becomes a *farm*stead.

The products that could be sold from the farm are a diverse assortment even more vast than the selection of corn varieties. You can sell vegetables, baked goods, bedding plants, eggs, broilers, field crops, small grains, beef, fruit, fish, trees, poultry, pork, timber, hay, horses, milk, herbs, sheep, nuts, grapes, flowers, honey, garlic, cider, and several other agricultural products. You know that you want to sell what you grow or raise on your farm, but the choice of *what* to sell gets rather complex.

"If you are striving for intensive production then it makes sense to rely on intensive marketing, too. Grow the higher value crops, bypass the middle people and look for ways to earn retail profits for your produce."

Andy Lee
Backyard Market Gardening

SMALL FARM PLANS WHICH OFFER DIVERSITY
. . . AND PROFIT!

APPLES AND PEARS
FEEDER PIGS
POTATOES
GEESE
BROILERS
GARLIC
RABBITS

MUSHROOMS
SHEEP
TURKEYS
BROCCOLI
CABBAGE
CAULIFLOWER
BRUSSELS SPROUTS

LAYING HENS
MARKET GARDEN
STRAWBERRIES
SHEEP
CLOVER-GRASS
FIELD CORN
RHUBARB

BLUEBERRIES
RASPBERRIES
MARKET GARDEN
SWEET CORN
GUINEAS
GREENHOUSE PLANTS
NUT TREES

Now back to that single kernel of corn. After an exhaustive search for what you want to end up with (the *end product*), you plant that kernel of corn. You apply soil amendments or compost to enrich the soil. Your site selection is in full sun, where the future plant will thrive. You cultivate the soil and ensure that it is free of competitive weeds. You work the soil to make it loose and friable. You then place the corn kernel at the correct depth, where it can be assured of moisture, the correct soil temperature and will germinate properly. You keep coming back to the planting spot every few days to check on the progress of the corn plant. Then one day you see a shoot of bright green emerging from the earth. Over the next few weeks, you will water the plant to ma̶ ̶ ̶re that it has enough moisture. You w̶ ̶ ̶ ̶e the plant to ensure that it has the c̶ ̶ ̶ ̶f nutrients. Again, you cultivate the ̶ ̶ ̶he plant to keep the weeds at bay a̶ ̶ ̶the soil from crusting. As the corn pl̶ ̶s, you defend it from pests by appl̶ ̶ ̶̶me sort of protection (chemical pesticide or a natural product approved for organic production). You watch the corn and measure its progress. If it is sweet corn, you know that when the plant is in silk, harvest is only three weeks away.

Manage, manage, manage. And then, one day, you have not only a single kernel of corn, but you have 200 (or 300 . . . or maybe even 400). The natural process of reproduction from a single kernel of corn is astonishing when you look at it in those terms. But even more exciting is the third generation, when you plant those 200 open-pollinated kernels

the following year . . . and they produce 40,000 kernels. The fourth year, you plant the 40,000 . . . and reap eight million kernels! It's astounding when you think about *the potential.*

Think about the following question for a minute. "Do you manage your farmers' market business as well as you manage that corn?"

Of all the methods you have available to sell your farm products, you have finally chosen farmers' marketing as your primary venue. *Whew!* By eliminating PYO, wholesaling to grocery stores, restaurant sales, a roadside stand, mail order, CSA, vertical integration, and conventional methods of mass marketing, it's nice to know that you can concentrate on the local farmers' market.

Business Management We All Can Understand

Have you chosen the correct site for your "kernel" by making sure that the market is well attended and popular? Have you prepared the "soil" by planning ahead (crop selection based not only on popularity, but value)? Have you monitored your early growth on a regular basis, checking for symptoms of potential problems? Have you frequently "watered" the business with scheduled financial investments, ensuring that the growth is not stunted by a sudden lack of "moisture?" Have you eliminated "weeds" that threaten your enterprise? Do you defend your farmers' market business from "pests" by close monitoring and taking preventive measures?

As the business grows, do you "cultivate" it on a regular basis?

Do you manage, manage, manage? If so, then your farmers' market enterprise can be just like a single kernel of corn. It won't only grow, but it can potentially be *very* lucrative.

Plan . . . and Dream

As the seasons pass by and you gain experience in marketing your farm products, don't spend all the profit every year. Reinvest a significant portion of your earnings.

Long before our family ever turned one spade of earth, we had a multi-year *plan*. The centerpiece of the plan was the kitchen of our newly acquired Shenandoah Valley farmhouse and three acres of land. No one saw this plan except for us. We constantly returned to the plan during the first two years to measure our progress. We put in new plans, modified some schemes, and quickly deleted some items when we found out there was little or no incentive (*read* $$$$) for producing the item.

As new opportunities presented themselves, we actively researched and sought advice from people who would know the answers (our customers primarily, and also successful small scale family farmers in the area). Of course, having Joel Salatin as a neighbor, friend, and mentor was a distinct advantage. When we learned about Pastured Poultry, we implemented it. When we learned about high value orcharding, we implemented that also. While we still haven't accomplished all of our goals, we have plenty to strive for. **The beauty of market gardening and farmers' marketing is the dynamic, ever-changing aspect**.

In retrospect, our original strategy turned out to be a sound plan. It gave us *direction*.

Our Own First Plan

A. CATERING/BAKERY BUSINESS

The goal of our home based catering business is to provide a significant family livelihood by offering freshly baked items such as cakes, breads, cinnamon rolls, etc. to the community. The initial thrust of this business, which we will name BITTERSWEET FARMSTEAD, will be to bake and sell goods out of our home's summer kitchen and the local farmers' market. We will expand into the larger market of full service catering, including business meetings, reunions, weddings, picnics, conventions, etc. Sheila's experience in this field gives us a definite advantage in operating this business, as evidenced by success. Her expertise lies in the areas of planning the event with the customer, coordinating the hired help, and actual preparation and presentation of the catered food items. Jeff's experience in business management, sales, maintaining records, advertising and market evaluation brings into the enterprise the required traits necessary to succeed. Each party complements the other.

1. FIRST YEAR

If the acquisition of the property at Bittersweet Lane becomes a reality, the first few months will be spent renovating the property's 100-year-old summer kitchen for use as a bakery and commercial kitchen. Tasks to get the venture started include:

- structural analysis of the summer kitchen

- gut and clean the interior

- renovate the building to accommodate a small commercial kitchen/bakery

- obtain Local Health Dept. & State Department of Agriculture Permits

- obtain business license

- open business checking account

- devise a simple but accurate accounting system

- make contact with foodstuff wholesalers and establish an account

- advertise by the most efficient and cost-effective means

While renovating the summer kitchen, it is important that we offer a small quantity of baked goods in the community via the farmer's market. This is in order to gain an initial customer base and establish a good reputation. Sheila and I may even drive door to door in the Middlebrook/Arbor Hill area and give away free pans of cinnamon rolls, quarters of Granny Cake, and sourdough rolls. This is in order to meet our new neighbors and let them know what we have in the works. We can also let them know that our products are available every Saturday at the farmers' market. These goods will be made in the home kitchen until the summer kitchen is fully renovated. During the first few months, we will concentrate on the following bakery products:

- Granny Cakes

- sourdough rolls

- cinnamon rolls

- sourdough bread

2. SECOND YEAR

After a year spent exploring and developing the market and renovating the summer kitchen, we will be able to focus more on the catering aspect of the business. This will be a direct "spin off" of the bakery, and we will accomplish the following:

- offer the community a full service catering business to include: brunches, business luncheons, weddings, reunions, church functions, etc.

- expand the line of baked goods

- *increase the outlets for the baked goods*

- *study the first year of operation and eliminate wasteful or unprofitable practices*

- *initiate procedures to bring the business to its full potential*

3. THIRD AND FUTURE YEARS

After two full years of establishing a market, perfecting our line of baked goods, and streamlining the business operations, we should be at a point where we can pause, and determine where we want to go with the business enterprise. Currently, our primary desires are:

1) to provide a stable and adequate family income
2) to allow us to work independently in a home-based business
3) to enjoy the pleasure of seeing our customers completely satisfied with what we are providing them.

B: MARKET GARDEN

The goal of our market garden is to provide a steady and reliable income for our family and to offer to the community fresh and healthy produce from a local source. The reasons we have elected this particular livelihood are:

1) to work independently and in a stress-free environment
2) to supply the market demand for chemical free produce
3) to complement the bakery business
4) it's a good clean living.

1. FIRST YEAR

During the first year of the market garden, the primary focus will be to prepare the soil for future fertility. The means of achieving this goal will be to work the soil of a one or two-acre area and plant continuous green manure crops in succession. The green manure crops will consist of buckwheat and annual rye grass. The initial garden (6000 SQ. FT) will be for home use and learning the Eliot Coleman method of organic growing. Learning the system and gaining local knowledge about cultivars will be the principal intent. Small stock will be maintained to both supply meat for the family table, and to contribute animal by-products (manure) to the market garden for soil fertility. Surplus produce will be sold at the farmer's market on Saturdays, fed to livestock, or given to friends.

Projects for the first year include:

- *have soil tested for mineral content, humus, and tilth. Test soil in immediate garden area, plus back two acres.*

- *lay out a 60' x 100' garden with 4' beds*

- chisel plow, break up and incorporate 35 - 50 tons of aged horse or cow manure per acre on back two acres during late spring. Also, add 1 ½ tons of colloidal phosphate per acre, and enough limestone to bring soil up to proper levels of pH. Green manure same two acres with buckwheat during summer, and plow down when it flowers. Sow annual rye for a winter cover crop.

- construct six 5' x 5' compost bins

- use a shredder to maximum advantage in creating a compost system

- build a small "A" frame pig shelter

- fence off two 30' x 30' pens for feeder pigs and chickens

- build a small corn crib (late summer)

- set up pig and chicken pens with feeders and waterers

- order 100 raspberry plants and start a berry patch

- start permanent asparagus and horseradish beds

- order 50 straight-run Barred Rocks, and 50 mixed cockerels from hatchery for May 17 delivery

- purchase two feeder pigs locally in May (from Mr. Glover)

- plant a 70' x 70' field corn plot (two weeks after sweet corn)

- plant a 70' x 70' feed patch with leftover vegetable seeds, grain sorghum, etc.

- plant a 50' x 50' gourd patch (minimum 500' away from garden)

- plant double row of sunflowers to shade pig and chicken pens

- purchase only one month supply of livestock feed at a time

- use a grain mill to maximum advantage in making chicken feed

- use chipper-shredder for making animal bedding out of straw, corn stalks, etc.

2. SECOND YEAR

The second year will be an expansion of the first. Without going full time into market gardening, the intent is to apply the knowledge that was gained during the first year and enlarge the garden area. Ample surplus should be available to sell at farmers' market on a regular basis.

Specific projects for the second year include:

- Move garden to previously green-manured plot and expand it to 120' x 100'. If this is not feasible, or if soil

19

on back two acres needs additional building, keep garden in the previous location and concentrate on intensive use of the Eliot Coleman method. Continue green manuring the future market plot and incorporating aged manure, minerals, etc. Don't rush the soil.

▸ *Rotate small stock pens in accordance with a long range plan (four pens . . . field corn > pigs > feed > chickens)*

▸ *raise two (maybe four) feeder pigs*

▸ *order 50 straight run Barred Plymouth Rocks from the hatchery. Keep these birds in the small pen rotation plan.*

▸ *Study the Hoop Coop method of ranging broilers, and if there is a viable market, order up to 300 White Rock cockerels and range them on the future market plot.*

▸ *expand 30' x 30' pens to 70' x 70' and fence in two remaining 70' x 70' pens. End result is four 70' x 70' pens for rotation*

▸ *add a simple hoop greenhouse for seed starting (12' x 20')*

▸ *move 50' x 50' gourd patch to opposite corner of property*

▸ *cultivate, mulch, and harvest first berry crop*

▸ *in autumn, break up and plant green manure in a small area for a feeder lamb paddock to be implemented the following year. If this space can be more effectively utilized as part of the market garden, cancel, or delay raising feeder lambs for market*

▸ *build up compost system for large quantities*

▸ *enlarge berry patch by 100%*

▸ *if feasible, plant a ½ acre to field corn*

▸ *concentrate on reducing the amount of commercial feed purchased by raising small plots of grain including grain sorghum, oats, etc.*

3. THIRD YEAR

The goal is to go full time with the market garden during the third year. By this point, enough knowledge will have been gained to expand garden size to two full acres. Specific tasks during the third year are:

▸ *expand to two full acres for intensive market gardening*

▸ *design and install a drip irrigation plan*

▸ *set up walk-in cooler (8' x 8' with straw insulation). Use existing concrete block storage shed if possible. Cool with a used 4000 BTU* room air conditioner or a salvaged refrigeration unit (keep costs low)

▶ build a washstand out of a salvaged bathtub. Use runoff for watering garden.

▶ enlarge berry patch to at least 300 plants (twelve 50' rows)

▶ increase size of laying flock to 75 hens for market eggs

▶ investigate feasibility of raising pheasants for market

▶ rotate small stock pens in accordance with long range plans

▶ purchase four feeder pigs

▶ purchase four feeder lambs

▶ fence in 150' x 150' paddock for lambs and plant to tender grass and legumes

▶ focus on primary market crops of garlic, pumpkins, potatoes, cut flowers, salad greens, peppers, etc.

▶ break up and green manure a third acre, if feasible, for expansion of market garden to three acres the following year

4. FOURTH YEAR

The chief objective is to grow slowly and increase the profit margin. Decreasing purchased inputs will raise profits. Study new ways of marketing produce if necessary. Take a long hard look at setting up a CSA (Community Supported Agriculture), permanent farm stand, or home delivery business. Consider selling compost, mulch, bagged manure, etc. at farmers' market. Fourth year tasks include:

▶ expand market garden to three acres

▶ rotate poultry to a new pen

▶ rotate small stock pens in accordance with long range plans

▶ continue with feeder pig program and expand as necessary

▶ if feasible, order 100 broiler chicks for range feeding and include in long range crop rotation plans (Eliot Coleman Hoop Coop Method). Investigate ranging the birds in raspberry patch.

▶ evaluate market and increase laying flock to 100 hens if demand is good

Don't try to get too big too fast. **Remember what you came here for.**

That was our original plan. As noted, it has been changed, modified, corrected, transposed and altered *numerous* times. But we *had a plan*.

Cold, hard *facts* we learned in our particular situation included:

❶ forge ahead with the laying hens because the market for eggs was excellent.

❷ scrap the idea of salvaging the summer kitchen. We discovered the local government inspection process would allow us to use our family kitchen in the home, and it was much more inexpensive to modify it than an entire outbuilding.

❸ the Hoop Coop method wasn't nearly as efficient and profitable as the Pastured Poultry model.

Briefly, what I learned about *planning* was this:

❶ It rarely goes as fast as you had hoped. Keep a critical eye on the cash flow of the operation.

❷ You can make-do longer than you thought possible. Living meagerly is not a lot of fun, but it certainly aids in building character.

❸ If something doesn't work, ditch it until you can get back to it later. Be flexible. Hardheaded farmers are broke farmers.

Your farmers' market operation should have a plan also (in writing!).

Intensive Effort, Intensive Profit

Your crop plan should concentrate on production - specifically *intensive* production. For the small scale farmer, intensive methods of farming are probably the only way to make a consistent profit.

What *is* intensive? In my mind, "intensive" applies to management methods and philosophies, not simply maximizing production at any cost on a given piece of earth during a single season. Intensive farming emphasizes:

▸ succession cropping

▸ season extension structures

▸ long term soil fertility

▸ high value crops

▸ inter-planting

▸ direct marketing

▸ efficiency

Intensive farming means optimizing local conditions to get the best yield possible from a small plot of land. I'm not so sure that you can intensively farm 800 acres. At least I haven't *seen* it done on that scale. But you *can* do it on a half acre . . . or three acres . . . or twenty acres. Intensive farming and regional food production go hand in hand. It is possible to feed dozens of families and make a handsome profit at the same time.

Plan to invest in your farmers' market enterprise on a regular basis. Intensive farming and marketing often require an intensive financial plan. If you have reached the physical limits of one slot at the farmers' market, expand to two. If you still have surplus produce, attend another market or add another day to your marketing week. This will require additional capital and labor, but the results will be worth the effort.

The best books around concerning intensive methods of crop production are Eliot Coleman's *The New Organic Grower*, Dick Raymond's *Joy of Gardening*, and Andy Lee's *Backyard Market Gardening*. When it comes to humane, healthy and profitable livestock production, Joel Salatin's *Salad Bar Beef* and *Pastured Poultry Profit$: Net $25,000 in 6 Months on 20 Acres* are the best books available for the direct marketing family farmer. A dynamic farmers' marketer can learn a lot about production techniques, planning, and *success* from these four gentlemen.

Lesson learned by Mr. Jim Chaffins at the Staunton/Augusta Farmers' Market in Staunton, VA. His customers hesitated at paying $4.00 per pound for fresh garlic . . . so he changed the packaging and the price to 50¢ an *ounce*. Now he regularly moves *more* garlic . . . at a *higher* price!

Chapter 4

Market Organization

T he very first thing we should understand is the punctuation in the following term:

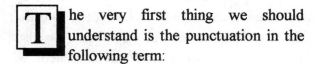

Farmers' Market

It is *correctly* placed. The apostrophe is used to express possession. If we omitted the apostrophe, then the term would just mean a group of farmers (plural) and a market (singular). That situation is common in modern agriculture: a lot of farmers . . . and a market that shows no *direct* relationship with farming. If we moved the apostrophe to the left by one space, then we are using the possessive for a singular term. Now we have one farmer . . . that *owns* the market. By placing the apostrophe where it is, however, we are expressing the idea that it is the *group* (plural) of farmers who possess the market. It implies a mutual association for the benefit of many farmers. Make sure that your farmers' market

has the apostrophe in the correct position.

Let's look again at how the USDA defines a farmers' market:

> **"A common facility or area where several farmers/growers gather on a regular, recurring basis to sell a variety of fresh fruits and vegetables and other farm products directly to consumers."**

This is an accurate and respectable definition. Here's why:

Market Defined

It uses the phrase "common facility or area," which implies a public location and not an exclusive business establishment. It implies an association of farmers rather than a single private enterprise.

The definition says that a farmers' market is "regular" and "recurring." Indeed, your operation should be both. An irregular or occasional market is confusing to the public. They want to know that if they get out of bed at 6:30 A.M. on a Saturday morning to get fresh tomatoes, a dozen eggs, and a pound cake straight from the farmer, the market will be there *and* open.

What is sold by this definition? "Fresh fruits, vegetables, and other farm products." I like it. This means that a farmers' market is not limited to green beans, apples and sweet corn. It tells *me* that we should be able to sell mulch straw by the bale to city gardeners; fresh eggs from the farm; a pound of sausage from our own hogs; a gallon of milk from our Jersey cow; and a loaf of whole wheat bread from our farmhouse kitchen. Do you think that this is what the customer *wants* . . . fresh food from the source? *You'd better believe it*!

Finally, the definition states "directly to consumers." Direct sales methods provide many small and medium sized farmers with their only access to the public. Many small scale growers have been shut out from the mass markets of agribusiness by their inability to buy high-tech packaging systems, produce cantaloupes in February, and their inability to "measure up" to market standards. Farmers' markets are designed to sell locally produced food and farm products directly to the customer. It is the customer that sets the standard . . . and not a middleman. By choosing this method of selling my produce, I'm deliberately electing to sell to my neighbors in my hometown. Let the corporate conglomerates feed Russia, China, and Europe. They think they do it very well. As for me, I'll sell fresh and nutritious food to my friends and neighbors . . . right here . . . in my own community. And I'll make a profit at it.

A Look Back

Farmers' markets have been around a long, long time. For centuries, growers have been bringing in food from rural areas to sell to the townsfolk. Historically, these markets have been fresh-air, seasonal, and retail.

Most markets around the nation were originally formulated or organized by volunteers, to serve both the public and the farming communities. There are, however, also many markets that "just sort of happened."

Consider the following:

Two farmers in different vehicles are on their way to the local grocery store where they normally sell their produce. They stop at an empty parking lot to chat about the weather, the crops, and the prices that the grocer is "giving them."

While they lean against the front fender of one farmer's truck, a car pulls alongside. Two ladies get out of the car and walk over to the back of one truck to inspect the fresh tomatoes, lettuce, and peas. One of them says "Those are some nice looking tomatoes. Did you grow them yourself?" One farmer replies "Yes. Less than ten miles from here. Picked 'em this morning as a matter of fact." The lady nods her head, then asks "Your prices wouldn't be higher than the grocery store would they?" The farmer self-consciously responds "Uhhhhh, no." And then more confidently "No ma'am." He knows that at retail, he will receive 40% more than he expected.

Looking at the other farmer, light bulbs of ideas are racing through his mind like jolts of electricity. He likes what is happening here. He then returns his attention to this gentlewoman and says "No ma'am. I couldn't ask more than the grocery man does. My tomatoes are the same price as his tomatoes." She quickly says "Good. I'll take 20 pounds. My friend and I have a lot of canning to do this afternoon. How about those peas over there? They look pretty good also."

Her friend then chimes in by asking the other farmer "How about you? What have you got on your truck?" As the conversations continue, another car pulls up . . . and then another.

A farmers' market has just been born.

Fundamentals of Organization

Numerous farmers' markets have no rules at all. Relying on tradition and a mutual understanding, they simply "happen" every year at the same location. No one person is in charge and there is no formal organization. These markets are suitable for some locales, but the successful market eventually organizes and establishes some rules and guidelines. The group then becomes more efficient in its efforts.

In Appendix D, you will find actual market rules and guidelines that were collected from a diverse group of farmers' markets nationwide. Read and study these samples. Then apply what is best for your own situation.

Let's take a look at some fundamentals of farmers' market organization.

- ▶ Every market should have a market master. Someone must be the administrator, the person who is in charge. It can be a farmer, or it can be a third party.
- ▶ Every market should allow individual input from the farmers or vendors.
- ▶ All successful markets have written rules and guidelines.
- ▶ The written rules must encourage fairness and free-market opportunity. They should not create red tape for the farmer. The rules should *eliminate* obstacles . . . not generate them.
- ▶ All markets should have a committee, a council, or board of directors to administer the normal operations of the market. The people involved should be volunteers who are *passionate* about the market . . . they should not be government bureaucrats burdened with yet another *civic duty.*

Market Master

The market master has overall responsibility for seeing that things go well on market day. Duties of a market master should include:

27

- Making sure that the market area is ready to accept vendors.
- Ensuring market signs are up and legible.
- Assigning slots for the vendors.
- Resolving misunderstandings between vendors.
- Ensuring that the market commons is safe and prepared for customer foot traffic.
- Determining that all rules of the market are being heeded.
- Collecting market fees, sales taxes (if applicable), etc. at the end of the market day.
- Ensuring that the market area is cleaned at the end of the day and returned to its original condition.

As you can see, a Market Master has a very complex job, and it should not be taken lightly. The Market Master, if *not* a farmer or vendor, should receive compensation for carrying out these tasks. Even if the Market Master is a participating farmer, perhaps it would be appropriate for the organization to waive the normal space fees paid by the farmer in return for accepting these all-important responsibilities. The compensation of a Market Master should be commensurate with the responsibilities, the proficiency, and the management skills of the person selected. Larger markets have a full-time Market Master with a salaried rather than an hourly wage position. Smaller markets hire someone part-time who is not only interested in the market, but will become an integral part of the effort.

Selection of a suitable Market Master is *critical* in the success of a market. The person should be pleasant, dynamic, a problem-solver, and someone who pays attention to the details. A *good* farmers' market has a *good* Market Master. You can count on it.

The Market Committee

The committee, counsel, or board of directors should be comprised of people intrinsically involved with the success of the market. Most of these people will be self evident to the organizational group, and the appointment process should be painless for most. Seek people who are volunteers, and who are passionate about agriculture, regional food production, and the promotion of the local community.

A diverse and dynamic group may include retired farmers, farm store merchants, Directors of Tourism, local restaurant management and chefs, Farm Bureau members, and Cooperative Extension agents. You may be able to enlist a member of the local media to join the committee. Media folks usually appreciate being asked to be a part of the community effort, and can normally offer valuable advice on public relations and market promotion. In addition, you have a built-in correspondent to help with the PR effort. You might also ask a college student to become a member of the committee. Young adulthood has its own level of energy, and an interested, vibrant young student can provide critical input to show the "old-timers" what is necessary to attract younger customers to the market.

The committee should also have one or two farmers' representatives or vendor delegate to act as an integral part of the group. This person can provide essential feedback from the farmers and vendor on major decisions. This person should also be "the voice of the farmers" and not an independent agent with only his or her own operation in mind. The selection of a farmers' representative or delegate can be accomplished by either asking for a volunteer, or by nominations and the popular vote of the farmers and vendors.

Responsibilities of the committee, counsel, or board of directors should include:

- ☛ Writing and implementing a set of market rules and guidelines.
- ☛ Resolving any issues that are beyond the scope or ability of the Market Master.
- ☛ Establishing both short-term and long-term goals of the farmers' market.
- ☛ Instituting an annual budget.
- ☛ Arranging advertising and promotions for the market.
- ☛ Meeting on a regular basis to manage the usual business of the market.

Equitable Market Rules

As discussed before, let us reestablish that this is a *farmers' market* and not a craft exhibition, a yard sale, or an antique show. The customer comes to the farmers' market wanting *farm* products.

Read the last sentence of the previous paragraph again for good measure.

Now that we agree on what a farmers' market is, let's take a look at some very sound and objective rules. They may not apply to every market, but they make sense for most.

The majority of farmers' markets across the country have restrictions on who can sell and what can be sold. These rules are usually at the mutual agreement of the participants.

Over half report that farmers can only sell their own crops and farm products.

When it comes to restrictions, "Producers Only" markets are in the majority.	
Sell only crops that you produce	53%
Other restrictions on selling	26%
No restrictions on selling	21%

About one fourth report that they have *other* restrictions on selling, which may include:

- ▶ Farmers must not resell another farmer's produce

- Farmers may sell a neighboring farmer's produce, but the neighboring farm must be within three miles of the participating farmer
- Artists and craftsmen are allowed during slow periods, at the beginning and end of the market season only
- Items such as cheeses, natural syrups, honey and cider must be produced, collected, or processed by the vendor
- Processed foods must be made from a majority of locally grown products
- No cut flower sales due to pressure from local florists

Each market takes on a personality of its own as rules and regulations are developed. Some farmers and vendors will gripe and carry on, and they may even move on to other markets in the area or choose another method of selling their goods. The market committee must realize that this will occur sporadically. But its decisions are critical in setting the direction for the market. Effectively implementing the rules and enforcing them without compromise will allow those goals to be achieved.

A few states classify farmers' markets by their selling methods. A prime example is one state that distinguishes:

- **Class A markets**- farmers can only sell what they grow themselves.
- **Class B markets** - a bona fide farmer can sell what he or she grows plus produce from one other local farm. The definition of "local" is determined by the Market Committee.
- **Class C markets** - a bona fide farmer can sell for any number of additional bona fide farmers, provided the produce is labeled as to its origin.

New farmers' markets usually have the goal of being a "producers only" market, or in the case stated above, a "Class A" market. A *few* have successfully started off with this regulation. Most, however, allow a wide variety of reselling *during the first season*. The intention is to get the market on its feet by drawing in a lot of vendors, an abundance of customers, and plenty of attention. When the market is stable and has a steady core of farmers and vendors, most markets choose to gradually mature into a "producers only" market. By the third season, everyone should know if the market can accomplish this or not. It all depends on customer demand. The market committee should be out there on the sidewalk talking with the buyers. They should ask these customers what *they* want in the market . . . and then deliver.

When forming a new market, ask a local attorney, or even a judge, to address the market committee as to legalities and selling restrictions. One market learned the hard way (via a lawsuit) that it could not restrict its selling to only vendors from within state or political boundaries. It could, however, specify that growers be within a certain radius of miles. This evidently did not interfere with laws addressing interstate commerce.

Research the Law . . . You May be Surprised

By the way, many farmers believe that it is next to impossible to retail your own meat (beef, pork, poultry) at a farmers' market. There are a multitude of governmental regulations about health concerns and the selling of perishable items. Some of the laws are good . . . and some are awful! Don't let these sometimes ambiguous regulations be a hindrance to you. Research the law and sell your own sausage, your own steaks, and a gallon of milk from your own cow. In many cases, it *is* possible, and it is *exceptionally* profitable!

So . . . *do you allow* a truckload of watermelons to be sold at your rural Vermont farmers' market in May? If you want to be a "producers only" market, probably not. Before making a decision, however, get legal advice.

After you get professional advice on selling restrictions, rely on good old common sense. The basic premise of your effort should be that farmers' markets promote regional food production, benefit local family farms, and offer fresh and wholesome products to the community.

If you establish those premises, organize your market thoughtfully and skillfully, and *keep the apostrophe in the right place*, you will have a successful and dynamic farmers' market!

A well organized market promotes a pleasant experience for both
the vendor and the customer. The commodity and size of operation are
unrelated to the pleasure perceived.

Chapter 5

Vegetables that Customers *Want*

How many times have you made a run to the grocery store to buy a specific item and found an empty shelf? Back in the car and on to store number two. Again, an empty shelf. Zoom, zoom, zoom, to store number three, where, *finally*, there it is on the shelf just as pretty as you please - exactly what you were looking for. A big smile comes across your face, you tell the cashier of your trials as the money slips from your wallet, and bingo . . . everybody is happy.

The entire focus of the retail grocery industry is to supply the customer with what

Fresh vegetables have always been the most demanded products at farmers' markets.

he or she wants, and do it at a profit. Your focus as a dynamic farmers' marketer is exactly the same. Don't forget that.

Of course, what the customer requires at the grocery store and what they desire at the farmers' market are two *entirely* different things. A 20,000 square foot retail grocery can supply a huge variety of consumer goods. Not only food, but motor oil, a copy of The National Enquirer, laundry detergent, a bottle of wine, a greeting card, and a prescription refill. We farmers can't compete with that. We just cannot supply people with *all* of their necessities. We do, however, have the market cornered on one thing . . . *fresh food*. That, my friend, is primarily what people come to the farmers' market for. If you can produce a

Based on my observations, the most sought after food items at any farmers' market are (in order of descending volume) fresh vegetables, fruits, breads, plants, eggs, dairy products, meat, and herbs.

Of course, there are also goods that the customer wants *besides* food at the market. But right now, let's focus on what *you can sell* that the *customer can eat*. In this chapter, we'll review the vegetables that customers want.

Popular Veggies

The most common farmers' market vegetables are listed below:

Asparagus	Cucumbers	Pumpkins
Beans	Eggplant	Radishes
Beets	Garlic	Shallots
Broccoli	Kale	Spinach
Brussels Sprouts	Leeks	Squash, Summer
Cabbage	Lettuce	Squash, Winter
Carrots	Okra	Sweet Potatoes
Cauliflower	Onions	Tomatoes
Celery	Peas	Turnips
Collards	Peppers	
Corn	Potatoes	

**COMMON VEGETABLES GROWN
FOR FARMERS' MARKETS**

steady supply of high quality and *fresh* vegetables, fruit, breads, and meat, then you will be a successful farmers' marketer.

Obviously, you may not be able to grow all of these crops for the market. Even if you could, you probably would not want to. Based on your own situation, you have to select the crops that are highest in demand, produce the best profit, and fit into your growing season.

To give you an idea of what is in highest *demand*, the USDA reports the following:

1. Potatoes	11. Peas
2. Tomatoes	12. Brassicas
3. Lettuce	13. Spinach
4. Corn	14. Beets
5. Onions	15. Peppers
6. Cucumbers	16. Squash
7. Cabbage	17. Asparagus
8. Beans	18. Chard
9. Carrots	
10. Celery	

THE TOP EIGHTEEN VEGETABLES IN QUANTITY CONSUMED

Looking at the chart above, it is easy to discern that one of the reasons the top items are where they are is because they complement fast food. What does almost every American get at the hamburger stand to go *with* that quarter-pound burger? French fried *potatoes,* of course! And what indeed goes *on* the burger? *Tomatoes, lettuce, and onions.*

These top five vegetables will sell well at any market, but can *you* (small-scale producer that you are when compared to the corporate fast food giants) make a profit from these common vegetables? Yes, you can! Are you in the same ball park with the big boys? In some ways, *yes you are*!

The advantages of a small-scale vegetable grower over a conglomerate grower are usually:

- higher quality
- fresher from the field
- lower transportation costs

Whereas the major producers are looking at costs by the acre, *most* small-scale farmers need to look at costs by the square foot.

Square Foot Economics

For example, sweet corn is definitely a high demand crop; but is it a high profit crop for the grower with limited land? When we get $4.00 a dozen for sweet corn, that's about 33 cents per ear. If we get one marketable ear per three square feet in our scale of production, then a square foot of land has grossed 11 cents for the season.

Now what if we planted snow peas on that same square foot in spring and harvested a quarter pound of peas from it? At market prices of $3.00 per pound, we've grossed 75 cents from that square foot. A week later, that same square foot will generate another picking . . . another quarter pound of snow peas . . . and another 75 cents.

At the end of the pea harvest, we work the soil and sow radishes. Twenty-one days later, we harvest the radishes. They produce two bunches per square foot. We sell the radishes at the market for 50 cents a bunch. Add another buck to the earnings for that same square foot.

Over the next month or so, we plant

buckwheat for a cover crop, and let it mature to the bloom stage. Then we till the buckwheat under to incorporate more organic matter into the soil. A few days later, we set broccoli transplants for the fall season. It takes two square feet of broccoli to produce one head (which garner $1.50 each at the market). Our square foot of broccoli produces yet another 75 cents.

Let's add it up. We grossed $1.50 for the peas and got $1.00 more for the radishes. The soil (and our honey bees) benefitted from the buckwheat crop. Then we got another 75 cents for the broccoli. I've got $3.25 in my pocket now rather than 11 cents for sweet corn. Sure, it took a lot more labor, and it took more investment (seed and transplants), but the results were *worth* the effort. Five or six thousand square feet of this type of intensive cropping, and you are looking at some potentially lucrative income . . . and this scale of production is *not* a full-time proposition.

Why grow sweet corn?

Because people want it, that's why. If you grow vegetables on more than an acre, you *should* grow some sweet corn for market. Smaller market gardeners, however, should probably stay away from it. You can simply make more money by concentrating on the higher *value* crops.

There *are* some exceptions to this suggestion, however. You, as a dynamic farmers' marketer, must take advantage of the market when the timing is right. Many farmers

- Asparagus
- Bush green beans
- Broccoli
- Carrots
- Edible pod peas
- Green Bunching Onions
- Head lettuce
- Leaf lettuce
- Sweet peppers
- Pickling cukes
- Spinach
- Tomatoes

TOP TWELVE VEGETABLES IN ECONOMIC VALUE *Listed in alphabetical order. Values based on pounds produced per square foot, retail value per pound at harvest time, and length of time in production.*

shy away from growing potatoes, for instance, because it is considered by some a large acreage crop that is potentially volatile in conventional marketing methods. By selling customer-direct with your farmers' market operation however, that just isn't a factor. One summer, I read reports of potato farmers in Idaho getting only $3.00 per cwt. (hundred pounds weight), and there were rumors of $2.00 per hundred. That's only 3 cents per pound! *Nobody* can grow anything for that price and come out with his shirt on! The *same* summer, we were getting 50 cents a pound for fresh, local potatoes. A vendor at a nearby market was getting 60 cents for some of his exotic varieties. Why? It's all in the marketing! When you take out 3 to 6 middlemen, *then you can make some money* on vegetables (even common vegetables like potatoes) grown on your farm.

Let's analyze five of the top vegetables in *economic value* (listed in no particular order).

Tomatoes

Customers *want* fresh tomatoes. And, *yes*, you've got to grow them if you want to supply one of the most demanded vegetables at a farmers' market.

Can you make money on tomatoes? Positively. But as in all of farming (and most of marketing) . . . *timing is everything*. If you can have fresh vine-ripened tomatoes at the market before anyone else, you can almost name your price. With the correct display emphasizing fresh, locally produced, vine-ripened, sumptuous, mouth watering tomatoes (get my drift?), a dynamic farmers' marketer can have a very successful day. Two dollars a pound and people will form a line.

If you have fresh baked goods from your farm, and a crop of leaf lettuce on hand, you are in business. Place your early tomatoes between a loaf of sourdough bread and a basket of lettuce. Then place a small hand-written sign next to the display that reads **"Tomato sandwich for lunch?"** Stand back and prepare to man the cash box! It will be self-evident that the power of suggestion is in fact truly powerful. You'll make more money in *one week* with those *early* tomatoes than you will the rest of the season with main season varieties. It's tricky to grow them (and sometimes a risky crop), but when it pays off, it pays off big.

At the beginning of the season, sell your tomatoes by the pound. Later in the season, when demand drops off, start offering your tomatoes in bulk. Right next to your $1.00 per pound main season tomatoes, offer the customer (through signs) a five pound bag for only $3.00. Place empty bags next to the display and let the customer select the best

37

Tomatoes should be a staple for most market vendors.

tomatoes. When they've weighed five pounds out on the scale, pick out a nice tomato yourself, and toss it in as a gratuity. The customer will be satisfied, and you'll be moving a lot of main season tomatoes with this "win-win" technique.

It is common for the price of fresh, local tomatoes to drop 50% in just two weeks at the beginning of the season, but that should not deter you from growing tomatoes all year long. You might only break even with tomatoes in July and August when every home gardener has a vine or two in the back yard. Indeed, I've seen farmers practically *give away* tomatoes in the peak of the season just to get rid of them (Step right up ladies and gents! Ten pounds of tomatoes for only one dollar. Gotta move 'em today!).

Hey! If you have to give them away, do it. But, continue to supply tomatoes throughout the season. They are like a magnet for your market stall.

If you grow heirloom varieties, make sure people know it. Use the words "heirloom" and "antique" in your signs (you *do* have signs for every variety, *don't* you?). Grow several varieties including paste tomatoes, cherry tomatoes, slicing tomatoes, etc. Don't forget to bring some green tomatoes with you to the market also. With a little sign that reads **"Fried green tomatoes for breakfast on Sunday"**, I've seen one fellow move more green tomatoes than ripened ones . . . and at the same price!

Per square foot, tomatoes can be very profitable. Small scale producers can expect anywhere from 125 pounds to 450 pounds per hundred row feet, but intensive methods have yielded over 25 tons per acre! Before the season starts, formulate a plan for making some money with tomatoes. Then, follow the plan!

Pickling Cucumbers

Daily, cucumbers are shipped out of California, Texas and Florida by the truckload. Most of them are slicing or salad cucumbers bound for grocery stores, wholesalers, and restaurants via the conventional distribution system. There are thousands of people across the nation, however, who still like to make their own pickles. Many of these customers use recipes that are generations old and are very selective when it comes to choosing the cucumber for their family recipe pickles. They know that the only cucumber suited for making those delicious, crunchy, home-made, dill pickles are short, bright emerald green, fresh, pickling varieties normally found at farmers' markets. The white-spined types are especially popular with home canners because the end product seems to stay a vibrant white/green rather than turning yellow.

Most pickling cucumbers for the farmers' market trade are harvested by hand. Many market gardeners grow their pickling cukes on a trellis, which allows easier harvesting and also keeps the fruit straighter. The fruits are always selected by size rather than maturity. Premium pickling cukes should be harvested between 2 to 4 inches in length, and never allowed to get over 5 inches long. The ideal length to diameter ratio is 3:1 or less (slicing cukes are allowed to get much longer). You should grade your pickling cukes by size, with smaller, high quality fruits priced notably higher than the standard sizes. **Premium food should be premium priced**. Remember this important marketing principle.

Pickling cukes yield prolifically over a long period of time and do not take a lot of space if trellised. These aspects (besides the demand) are primary reasons that it is considered a high value crop. Making a profit with pickling cukes does not come without a large amount of labor, however. For the best quality pickling cukes, they must be harvested *every day*. A grower with staggered plantings (for a continuous supply throughout the season) will need plenty of help out in the cucumber field harvesting the fruits before they get too large. There are also several insects that just love cucumbers. Pest control via crop rotation, sprays, row covers, etc. is critical.

While pickling cukes are typically the highest value cucumber you can grow for the farmers' market, you should also produce some standard slicing cucumbers to remain diversified. Indeed, there is yet another variety of cucumber that you might also take a look at for high value. The extra long Oriental and European cucumbers typically grow from 15 to 24 inches long and are very popular in ethnic markets and in the restaurant trade. A short row of these cucumbers trellised in a small market garden can be very profitable. Provide recipe or serving suggestion cards for this unusual vegetable and *ask your customers*

what they think. You might find a nice little niche with these cucumbers that sometimes reach two feet in length.

Lettuce

Salad crops are popular everywhere in this health conscious age, and lettuce is the mainstay of all salad crops. Looking at it from the viewpoint of the customer, lettuce is one of the few crops that he can pick out from a wide variety of vegetables at the market, take home and *immediately* prepare a light, nutritious lunch within minutes. No cooking, no dirty pots and pans, and very little effort to prepare. It is an ideal purchase for most farmers' market customers.

Even though leaf lettuce is one of the most popular home grown garden crops in the United States, there is still a high demand for it at markets. Pale and watery iceberg lettuce found at the grocery store can often seem boring, and many people are aware of its low nutritional levels when compared to bright green leaf lettuce. Remember, one of the primary reasons for people coming to a farmers' market is to find something different, something fresher, something a bit more exotic than that found in the air-conditioned, fluorescent-lighted, chrome grocery cart crowded supermarket.

Butterhead and loose leaf lettuce varieties are excellent choices for selling at the farmers' market. Romaine is very popular, as are the fancy European varieties. Perhaps the biggest rage in recent years for salad crops is the introduction of mesclun mixes. These various mixtures of lettuce, chicory, cress, mizuna, mache, chervil, etc. are bringing premium prices to farmers who stay *aware of culinary trends*. Successful lettuce growers say that it is important to try and stay ahead of other producers by reading not only farming and gardening publications, but also reading vegetarian and health-related magazines. Doing what no one else is doing is a key in providing fresh greens for the lucrative salad market.

One of the most important things you can do is be consistent with your supply of fresh lettuce at the market. Customers who discover your delicious, nutritious, fresh lettuce will want to come back in a week or two and get more. You should be sowing seed *every week* of the season, utilizing a broad variety of heat tolerant and adapted cultivars. The high dollar return per square foot demands it.

Peppers

Bell peppers are a fine crop to grow for the farmers' market. They can be *very* profitable. Hot peppers and chili peppers sell well also (especially to diverse ethnic areas), but the sweet bell pepper is always in demand by the general public. I classify sweet bell peppers as any pepper that is not hot.

Several positive aspects about bell peppers include:
- they require little work
- they are fairly disease and insect resistant
- a pepper is simple to harvest

- they hold well on the plant

Commercial pepper growers average anywhere from 8,000 - 12,000 pounds per acre. Intensive methods can yield over 20,000 pounds per acre. On a smaller scale, you can expect at least 50 pounds per hundred feet of row. Keep your peppers picked when they reach harvestable size and color. This encourages more fruit to set, and, therefore, a more efficient harvest season. Red varieties can be temperamental when transitioning from green to red, so keep an eye on them and manage intensively. Red and yellow types usually command a premium price.

Normally, peppers are retailed at the farmers' market not by the pound, but by the unit - or in groups. A common price is 3/$1.00, 5/$3.00, etc. for peppers in season. At the beginning or end of the season, I've seen well-displayed, large, clean, fresh peppers bring $1.00 apiece!

Display your peppers in wooden bushel or peck baskets slightly tipped over to give the appearance of bounty. Always leave a portion of the stem on. Try using a handwritten sign that reads **"Had stuffed peppers lately? How about tomorrow with Sunday Dinner!"** If they are grown without pesticides and are chemical free, by all means *let the customer know*.

Asparagus

This is one of the most profitable crops you can grow for the farmers' market because:
- A single planting can last 15 - 20 years

- ▸ Demand is usually high
- ▸ It is easy to harvest and holds well in short term storage
- ▸ It requires minimal packaging (a rubber band)

Even though it only produces for a short time in the spring and early summer, asparagus is certainly a high-value crop for most market gardeners. The planting effort is massive, requiring 8-16" deep trenches with lots of incorporated organic matter. The prospects, however, for a long term stand are tremendous.

Asparagus can be planted from seed or by the crown. You gain a year by planting the crowns, but if you are thinking long term (and you should be) sowing asparagus by seed is extremely economical.

The average commercial yield is between 1500 and 2000 pounds per acre, and intensive methods have produced over 5000 pounds per acre. For the small scale grower, anticipate about 30 - 40 pounds per hundred row feet. Retail prices vary between $1.00 and $3.00, based on availability and time of the season.

Asparagus requires a large investment in labor, crowns, and time. Growers must patiently wait until the third year when the plants are mature before taking any harvest at all. But because the harvest season is 8 -10 weeks long for a mature stand, the potential payoff is well worth the effort.

Summary

The key to growing profitable vegetable crops is to think in terms of:
- ▸ pounds produced per square foot
- ▸ the retail value of the crop at harvest
- ▸ the length of time the crop is in production.

Good product + Good signage = Success at the Farmers' Market!

Chapter 6

Customers Will Stand in Line for Good Fruit

A fter fresh vegetables, it is fruit that brings customers to the farmers' market most often. Just let word get out during the week that the first strawberries of the season have been harvested and will be available at the local market on Saturday morning. Almost certainly, a crowd will be on hand at opening hour filled with the anticipation of buying those delicious, sweet, red orbs of fruit.

While many crops you may decide to produce for the local farmers' market are not "sure things," the possibilities grow geometrically when you elect to concentrate on fruit. Chemical-free and freshly harvested local fruits, especially, have many benefits as a profitable addition to your farmers' market enterprise.

The prices paid by consumers for pristine crops of geographically distinct fruits are almost absurd when you really think about it. You might ponder "Why would anyone in his right mind want to pay $3.00 or 4.00 for a modest serving (pint) of red raspberries?" The answer is "Because there is no other sensation in this world like placing a fresh, tart, juicy, lush raspberry in your mouth. It is to savor one of the most delectable tastes in the world, an indulgence that is seemingly beyond cost consideration." Fresh fruit seems to have a *luxuriant* appeal to some people, and they will pay dearly for it.

I think that both large *and* small fruits are deserving of your consideration when making crop selections for your farmers' market operation.

Large Fruits

It has been said that a single acre of the best fruit trees, well-managed, will produce more profit than a three hundred-acre farm, without a tenth of the capital invested. It is no wonder, then, that fruit production is so attractive for the small-scale, intensive producer.

You must be patient, however. The large

fruits, such as apples, peaches, pears and citrus take a few years to get in production. This factor is one that is critical in your decision making process. Many small farmers growing for the local market need something that will pay in a year or two. Lack of cash flow during this period must be considered. The long-term profits, however, of a well-managed orchard are considerable.

The market for large fruits is usually very stable, and the holding period (especially for apples) is a positive aspect. You need not feel so rushed to get the produce to the market before it starts to suffer loss of quality.

Apples

Probably the most reliable and popular large fruit in North America is the apple. U.S. apple production increased nearly 50 percent from the mid-seventies to the mid-eighties, and *fresh* apple consumption currently accounts for almost 60% of annual production.

A small apple orchard is particularly well suited to the northern climes because of the fruit's inherent requirement for a certain minimum number of days of cool weather. In modern times, however, warm climate varieties have been developed which offer growers below the Mason-Dixon line a chance to grow this most popular large fruit.

The best states for apple production are Washington, New York, Michigan, California, Pennsylvania, Virginia, North Carolina and West Virginia. Small-scale orchards in other areas, however, are able to produce an ample crop for the farmers' market by utilizing locally adapted varieties.

It usually takes three years from planting

apple trees before you can harvest the first crop. During the first two years, make sure that any fruit produced is removed from the tree to concentrate the plant's energy into leaf and limb development.

Be careful when selecting which varieties of apples to produce for your farmers' market operation. Apples bred for large scale commercial production are probably not a wise choice. Officials in the apple industry have admitted that their emphasis in breeding has been on color and size of the fruit. Taste, it seems, has taken a back seat. Backed by "extensive surveys," the megabucks apple industry has apparently concluded that color and size alone are what initially convince the general consumer to purchase an apple. The key word is "initially."

Is this the right approach for you? I don't think so. Your customers at the markets (who you know by first name, not via a survey) want *flavor*, and you should remember this when selecting varieties for your future orchard. A nice, big, ruby red apple is a beautiful thing to look at. If the flavor is unsatisfactory, however, the customer will not be back to purchase more of the fruit.

Apples are smitten with numerous pest and disease problems, but these can be overcome with proper management. Do extensive research, develop a management plan, and then *follow* the plan.

Some of the most popular varieties currently grown for direct marketing are Golden Delicious, Fuji, Braeburn, Arkansas Black, York, Jonathon, and Gala.

Display your apples at the farmers' market in peck or bushel baskets. Offer discounts for larger quantities. Try to keep the fruit cool and in the shade. Don't be hesitant to give away samples. If the market rules allow samples, keep three or four different varieties under a glass cover and slice off chunks upon request (*you* handle the knife, *not* the customer!). Sell only premium grade fruit. Wormy or bruised fruit must be sorted out meticulously.

Peaches

Peach production requires a high level of seasonal hand labor. This situation, in most cases, is ideally suited to the small-scale farmer and farmers' market producer. Most modest peach orchards offer the opportunity for the family to provide all of their own labor. Having to resort to hired hands from off the farm can cut into your profit margin substantially.

The popularity of peaches is revealed in the fact that more than five million bushels of the fruit are harvested *per month* during the summer season in the southeastern U.S. alone. All along the southern interstate highway system are huge signs that proclaim **"Fresh Peaches - Next Exit."**

Climatic conditions greatly influence the growing of peaches. The trees must have their "biological clocks" satisfied with a certain amount of wintertime chilling. This requirement varies with each cultivar. Popular varieties for small-acreage farmers include Redhaven, Reliance, Belle of Georgia and Elberta. Plant at least three different varieties which have successive ripening periods. This will stretch your season and decrease the all-at-once workload.

If not familiar with the crop, you should be aware that there are two different types of peaches to produce - freestone and clingstone. The freestone is usually the fresh market peach, while the clingstone is normally used for processing. In 1996, California produced about two-thirds of the nation's freestone (fresh market) peaches. This was due to the fact that spring frost virtually eliminated the entire crop in the "peach-belt" consisting of Georgia, South Carolina, and Alabama. A late frost is a peach grower's worst nightmare.

You should harvest your peaches for the farmers' market when the fruit is nearly mature, but still firm. A huge increase in both quality and size occurs during the last few days of ripening. You'll probably need to pick every two or three days during the height of the harvest. Handle the fruit carefully to avoid bruising.

Display your peaches utilizing the multi-level strategy, with at least three vertical tiers of fruit. The appearance of bounty is one that is most welcome when it comes to fresh peaches at the market! Sort and grade the peaches into quart, peck, or bushel baskets. You might tip the bushel baskets on edge, giving a "spilling effect" that accommodates closer inspection by the customer. Use your handwritten cards to tell the customer about the variety and what it is best suited for (fresh eating, canning, freezing, etc.).

Melons

Watermelons, honeydew melons, and cantaloupes are extremely popular commodities at farmers' markets. They can also be financially rewarding if you have ample ground to grow them on. Many vendors are concerned that their local growing conditions will not support melon production, but check again. There are a number of different varieties that have been developed in recent years which will grow in almost any region of the nation.

Melons are moderately labor intensive, and you need a substantial amount of acreage to justify a market scale of production. Less than a half-acre devoted to melon production would probably not be warranted unless you have a reliable and premium product with a guaranteed market demand (Disneyworld parking lot?). The space requirements are just too much for smaller market gardens.

The honeydew melon has the highest average sugar content of all melon types. Summer's sweet bonus for many people is to buy a fresh honeydew melon at the farmers' market on Saturday, then serve it to guests for Sunday brunch. The firm, pale green fruits will draw customers to your stand like a magnet.

The large varieties of commercial watermelons are typically grown in the deep south on vast acreage. For small-scale growers, look very closely at the unique varieties such as Sugarbaby, Crimson Sweet, and Amish Moon & Stars. These are not very prevalent in grocery stores, and customers will be willing to pay a bit more for these unusual melons. Yellow-meated and seedless types are possibilities also.

Cantaloupes, sometimes referred to as "muskmelons," are very popular with diet conscious consumers (who, these days, *isn't* diet conscious?). Dynamic farmers' marketers should capitalize on this trend. Cantaloupes are not difficult to grow, and premium fruits will command a premium price. Popular cultivars include Earliqueen, Jenny Lind and Athena.

If you sell at a tailgate market, leave the majority of your melons on the truck and let customers select from the pile. *Always*, however, have a sample melon cut open and visible from the front of your stand. You might cut cubes of melon for samples and keep them in a clear casserole dish surrounded by ice. Have a box of toothpicks handy for serving. If there is a slow spell at the market, keep yourself busy by washing and sorting the melons by hand. Keep only the best melons in front of the customer. Discard (or eat) the rest.

Small Fruits

The entire berry family is well suited to:
- small-scale production
- consumer-direct marketing techniques
- quick cash turnaround

The reasons that industrial agribusiness does not concentrate on these highly profitable crops include the immense amount of labor involved to harvest berries, the tender nature and short shelf-life, and the somewhat specialized equipment involved in production and harvest. Indeed, many of our nation's fruit and berry producers are just gardeners in a larger scale endeavor. The profit potential with small fruits even turns many a row-crop farmer into a small fruit producer.

Three of the most attractive and profitable small fruit market enterprises include strawberries, raspberries, and blueberries.

Blueberries

Blueberries are related to azaleas and camellias and are native to North America. One of the neat things about blueberries is that they will not ripen any further once picked!

More popular than ever, well over a thousand new products were introduced in 1996 alone containing blueberries.

A well-thought out planting of blueberries will include multiple varieties, as this fruit is one that will produce *throughout* the growing season. Early cultivars will start maturing in June or July, while the later ones will carry on through September. This certainly will keep market customers returning to your stall week after week, which is exactly what you want.

Seek out professional advice when it comes to variety selection. Check not only with the Cooperative Extension offices, but with other blueberry producers in the region. A half-day spent on a fellow producer's farm is a good idea and usually an education in itself. Unless your market area is saturated with blueberry producers, the fellow farmer will probably be proud to show you around his operation. Buy some of his berries, of course, while you are there.

Since most small plantings of blueberries will be harvested by the grower's family, it is crucial that ample time is allotted for handpicking. It takes a *long* time to pick several hundred pints of blueberries! Choose varieties that have long, loose fruit clusters. Firm berries handle better and keep longer than soft ones. The scar (point-of-attachment of the fruit stem to the berry) should be small and dry and should not tear when the berry is picked. Very dark berries may be unattractive, since they might appear overripe to some customers. Save these for your own eating or processing. Taste the berries before you start picking (it probably isn't necessary to twist your arm to do this!). Some blueberries are rather tart when they first turn blue, and may be a couple days away from being fully ripe.

Display your blueberries in pint or quart containers. This is one commodity that should have a clear cellophane wrapper over the top of the container secured with a rubber band to prevent spillage. You don't want blue sidewalks in front of your stall at the market. It will also keep customers from putting their fingers all over the fruit. Have a sampler container open and accessible for market shoppers.

Raspberries

Raspberries are always one of the most sought-after fruits at a farmers' market. This bramble berry is so well known for its luscious exceptional flavor, you will be fortunate if you can get your berries off the truck and onto your farmers' market table before they are snapped up by wide-eyed customers.

Raspberries are sensitive to handling and have a very short storage life, hence commercial production is rather limited. A small-scale producer that has an immediate market and doesn't mind intensive labor, however, can do *very well* with raspberries.

A half- acre planting is probably the upper limit for a family-sized operation with no outside labor hired during harvest season. A well-managed half-acre planting of red raspberries can produce more than 1500 pints. Retailing for $3.00 per pint at the farmers' market, it's tempting to plant more, but *don't plant more than you can pick.*

Make sure that your raspberry plantation includes both summer and fall producing varieties. This will give you two crops rather than a single one, and it will stretch the fervent labor requirement out over a longer period of time. Latham and Heritage are two of the most popular red cultivars, the former being widely adapted and very winter hardy. Heritage is an everbearing variety and if you mow the canes to the ground after the first hard frost, you sacrifice the summer crop; however, this technique almost guarantees you of having a much bigger (and possibly earlier) crop of huge, luscious, red raspberries in the fall.

Though purple or yellow varieties are readily enjoyed for their premium flavor, there is not a significant retail demand for them. In bright sunlight, the purple ones simply look overripe. It is possible, nevertheless, to educate and familiarize your regular customers with these distinctively colored raspberries and create a significant niche market for yourself.

Display your raspberries in ½ pint and pint containers. Clear plastic clamshells are available, but most vendors choose wooden or paper mache' baskets due to the favorable colors which compliment the berry display.

Strawberries

Strawberries are particularly attractive for farmers' marketers with restricted growing acreage. Fancy quality strawberries, indeed, will sell quicker than almost any other market commodity. *Large* sums of money are made by the vendors who are first at the local market with freshly harvested strawberries.

One of the prime favorites of summer, the strawberry may become a significant part of your market operation once you have experienced the pure exhilaration of having to call in extra help just to man the cash register! Successful vendors not only have strawberries for fresh eating, but suggest to their customers alternative uses for the fruit, such as making strawberry pie, homemade strawberry ice

cream, and, of course, strawberry shortcake. Numerous customers, however, find themselves munching on the sweet delicacies and discover that their strawberries are gone before they can even get them home!

Producing high-value cash crops on small acreage is a distinct advantage when growing strawberries. Many small farmers can harvest more than 10,000 quarts per acre, the retail value of which is phenomenal when compared to conventional crops. Production costs can be high, but the profit margin can be very appealing if you sell direct to your customers at the market.

Perhaps more than most crops, strawberries require an immense amount of skillful management. Choosing varieties that are best suited for your soil, your growing conditions, and your market are all critical to the success of the venture. Extending the harvest season by selecting early, mid-season and late varieties is an important factor to consider. You want to have those berries at your market stand *as often as possible*.

Display your strawberries in pint or quart baskets and make the display as big as you feasibly can. You want customers to see your strawberries from a distance, almost guaranteeing that your stand will be their first stop of the day.

Summary

A strong temptation for novice fruit growers is to think big and plant big. They frequently plant way more than they can handle. In most cases, it is wise to start small and generate immediate cash flow with other crops. Sizable orchards and berry plantations are a considerable initial investment in both time and money. You must be prepared to wait for the deferred, but potentially lucrative, return on investment.

Integrating fruits, both large and small, into your crop plan can be exceptionally profitable for your farmers' market operation. It requires a high level of management and a lot of intense labor. Those who fail to both learn *and* practice the techniques necessary to produce *high quality* fruit, unfortunately, will not realize a profit.

Learn your lessons well. Then, hand-tend your crops, cultivate meticulously, and concentrate on producing the tastiest and finest fruits that will command premium prices. Generous profits will be a result of your diligent labor. Nothing worthwhile comes cheap.

I encourage you to remember two things concerning fruit production for the farmers' market:
- Choice fruit sells the quickest.
- If it were easy, everybody would be doing it.

Add value to your products - turn fifty cents worth of raw product into a three dollar sale. Note multi-level display!

Tipped baskets are an excellent display tool for the Farmers' Market! The customer *must* be able to see the product!

Chapter 7

Baked Goods, Eggs, Meat, and Specialty Products

Along with fresh vegetables and fruits, dozens of different farm commodities are sold at regional farmers' markets. Among the most popular items: bedding plants, baked goods, canned goods, cut flowers, cheese, cider, eggs, honey, house plants, jams, jellies, and meats.

Some markets allow arts and crafts to be sold at the farmers' market; however, most prefer to keep the market as an outlet for *farm products* only. Traffic and sales may be unusually slow for regular vendors during the beginning or ending of a season. This *may* be a period during which your market committee considers inviting local artists or craftspeople to participate in a *single day* of the farmers' market. Ensure that everyone understands arts and crafts are not regular commodities at the market, but that the farmers have taken a special step in inviting fellow citizens to participate in the market for a special activity.

Value Added Products

One of the best ways to increase not only your gross income, but net profit at your market stand is to offer value-added products to your customers. Now "What is a value-added product?" you might ask. Well, value-added simply means that you take the raw product and further process or modify it in some way that the customer might. You save the customer from doing most of the work, but charge likewise for adding value.

An example might be the choice between selling fresh paste tomatoes and selling pint jars of pizza sauce. A pound of high quality paste tomatoes might gross $1.25; however, the same pound of paste tomatoes processed into "authentic, organic, farmer's pizza sauce" may retail for $3.75 per pint. There is much more work involved, but your gross sales have increased because you have added value to the original raw product.

Another example may be the use of some farm product that normally would not be of retail quality, but would be fine as a processed product. How about those sour cherries that you let go one day too long before picking?

Don't let them rot on the tree - pick them now and make cherry pie! Rather than taking a total loss on these cherries, you can work a bit harder and put a few more bucks in your pocket. A few more bucks here and a few more bucks there is what can make or break a farmers' market enterprise. Selling your products in this way adds much more value to your customer - and to you.

My wife never has to buy eggs for her baked goods. That's because we keep our own flock of pastured hens. This is yet another example of value-added. If you don't have to purchase the ingredients for a market product, then you are obviously way ahead. On one occasion we catered a *full meal* to a local Ruritan Club meeting and didn't have to purchase one single input! The bottom line on something like this is really nice.

"The first question to ask with a value added product is 'Will this product meet a specific need in the marketplace?' The gourmet field is crowded. Look for something that is unique. Make sure that what you are doing is different (not just better) than what anyone else is doing."

Eric Gibson
Sell What You Sow!

So don't just sell strawberries, sell strawberry jam. Don't just sell honey, sell honey lip balm. Don't just sell buckwheat, sell buckwheat flour. Add value to your raw products by thinking "Now what would my customer do with this?" Then *you* do it, and charge your customer a premium for the added value.

Let's look a bit closer at three of the most prevalent farmers' market products besides fresh fruit and vegetables.

Fresh Eggs

There is just *something* about fresh brown eggs on display at a farmers' market. Many customers return regularly to the market for this one particular commodity. You can watch the serious egg connoisseurs walking along the commons on market day with empty cartons in hand. They are not only recycling the cartons to "their" farmer, but coming in for a reload.

Is there any money in eggs? I answer with a resounding "Yes!" Whether you are gathering two dozen a day or 125 dozen a day, there is an excellent market for those farm fresh eggs.

To give you an idea of the scope of the commercial egg industry during one thirty day period, U.S. egg production totaled 6.43 *billion* during July 1996, *up* 5 percent from the 6.15 billion produced in 1995. Production included 5.46 billion table eggs and 973 million hatching eggs, of which 914 million were broiler-type and 59 million were layer-type. The total number of layers during July 1996 averaged 295 million, *up* 3 percent from the average number of layers during July 1995.

Most farmers' market egg retailers are on a small-scale of production that is more in tune with what customers typically perceive - a family sized farm with a small flock of healthy chickens running around the barnyard. Children gather the eggs twice a day, and mom and dad clean, grade and package the eggs for the farmers' market. A very wholesome scene indeed, and not far from how an small operation should be run.

If the average consumer had any idea of how commercial laying flocks are managed, believe me, you'd have more business at the farmers' market than you could handle. Commercial leghorn hens are tiny, frantic and upset at the slightest hint of motion. Their feet never touch ground, in keeping with industry's trend away from "floor birds" at every stage of the hen's life. The hens never see the light of day, taste a blade of fresh spring grass, or hear the crow of a rooster. *If* you were *allowed* to tour a commercial layer house, first you'd have to don a plastic "biosecurity" suit and shoes. The modern strains of production layers are sometimes so fragile that any microbial organism brought into the house might cause a total loss of the flock. You would walk down gloomy aisles stacked with cages on both sides stretching into a dusty ammonia haze farther than you could see. Cages are divided into five or six rows with four levels of cages per row. The hens are small, white, wistful-looking birds with intent dark eyes, large pale combs, and clipped beaks. There are sudden frantic movements and they crawl over and under each other inside the cage. Hens stacked at the bottom see only your feet.

It used to take 24 lbs. of food to feed 100 hens per day; now it takes just 22 lbs. The industry is working on developing strains of birds to lay the same number of eggs on even less food - yet another attempt to get something for nothing. Water, clearly, is much cheaper than feed. No wonder commercial

eggs are so runny and pale.

Is there any wonder that old-timers proclaim "eggs just aren't as good as they used to be?" They still remember the fresh, high quality barnyard eggs of their childhood. *This is what you have to supply.*

Choose a strain of chicken that is suited for health, livability, and production. The American breeds Rhode Island Red, White Rock, and Barred Rock are ideal for small flocks - and they'll produce plenty of nice brown eggs for your customers.

Make sure you take only fresh, clean eggs to the market, and be *passionate* about how your eggs are superior to typical supermarket eggs. Take pictures of your hens in the nest box and display the framed photographs at your market stall. Help the customer connect! Also, don't be afraid to price your eggs above supermarket prices. They are worth it - and you know it.

At the end of this chapter you will find a brochure we distribute to our market customers.

Baked Goods

Fresh baked goods from the farmhouse kitchen have an immense appeal to farmers' market customers. There is nothing so dramatic as the difference between an apple pie straight from the farmhouse and one that you pull out of the supermarket freezer. This is something you don't have to convince customers. Their sensory organs do all the selling for you!

Most markets and governmental jurisdictions require that baked goods come from an inspected kitchen. Don't let this scare you. As long as you only produce baked goods, you don't have anything to worry about. In our area, there is a definite line drawn by inspectors about those food items that can be produced for market and those that can't. When is the last time you heard of someone "getting a hold of a bad loaf of whole wheat bread?" It just doesn't happen. Whole wheat bread is safe, stable, and doesn't spoil easily. Cheesecake, however, is an entirely different product. It contains dairy ingredients and requires certain temperatures and packaging. It has a definite shelf life.

Contact your local health department or agriculture department about which baked goods you can produce for farmers' market retail sales. Breads, rolls, and cakes will probably fall under the jurisdiction of the agriculture department, while the more unstable items will be inspected by the health department.

In our experience, cakes and breads are two of the most popular baked goods. Try

---

"At times it seems very odd. I go to town with a pickup truck full of garlic and flowers and vegetables, and with an anxious heart. A few hours later I return with an empty truck and a sense of boundless promise - because I know that the exchange is still working, that my customers still want what I can grow and take to market . . ."

Stanley Crawford
A Garlic Testament

---

slicing your cakes in half so the customer can see the inside. If you have a unique cake that is rare and tasty, charge a premium for it. There are many customers that don't care how much a good cake costs. They are only concerned about getting it from the market to their dining room table, where they can invite neighbors and friends over to share the experience. They make that cake a centerpiece presentation, bragging "Ohhhh, you wouldn't believe what I found at the farmers' market this morning. There is this nice lady, the other farmers call her 'granny', who bakes these cakes at her farmhouse every Friday night, and then brings them to the market every Saturday morning. You have to get there really early though. She is usually sold out by eight o'clock in the morning." Duly impressed, they then indulge in a product that no grocery store can even come close to reproducing.

Hints about selling baked goods at the farmers' market include:

▸ Start with four or five primary items in your product line. It's better to have 12 each of five items than just one or two of thirty items.

▸ Introduce new items into your product line gradually, and give each item three or four weeks to "prove itself." Don't pull an item from the market after just one week. Give it time.

▸ *Tell the story* behind each baked product. People love a story. If it is an heirloom recipe, share the background. If the product is unique in any other way, talk about it. One enterprising young man at our market tells customers about the maple syrup donuts he sells by relating "Yeah. Dad and I tapped three maple trees back in February, and we boiled the sugar water down in big cast iron kettles. Then, mom helped out in the farmhouse kitchen when she . . ." *People buy into this*, and want to be a part of it. They tell your story to someone else, and this leads to even more customers.

▸ Label every single product with the ingredients. Also, include your farm name and phone number. Computers are wonderful for this sort of thing. *Always* label your baked goods, and make sure you get the ingredients correct!

▸ Use a multi-level display for your baked goods. Two or three black, steel, baker's racks are ideal. Tiers of old boards on apple crates work well also. Cover the boards with a tablecloth.

▸ Offer lots of small, individually wrapped baked goods that can be purchased and eaten on the spot. Examples are: muffins, scones, cookies, slices of pound cake, brownies, etc.

Meat Products

Many small farmers are producing meat products on the farm for their own family, but neglect to consider the idea of retailing poultry, beef, pork and other meat products to their customers at the local farmers' market. Rather than settling for whatever the situation is with conventional markets, those farmers could (and *should*) be getting *retail* dollars and making much more profit by cutting out most of the middle men.

Of course, the prevalent question is always "but what about inspection?" My response to that is "What *about* inspection? If the law requires it, then do it." It's a lot more effort, but you will eventually get paid for going to that trouble when the cash register starts singing. By taking your livestock to a state or federally inspected slaughter facility, you are doing what is necessary to satisfy the law. Check around first, however, with both the people who are in charge of meat inspection and the people who are actively involved in direct marketing their meat products. You may be pleasantly surprised that several meat products are exempt from inspection.

Let me share a story with you. Joel and Teresa Salatin, neighbors and friends of ours, arrive at the Staunton/Augusta Farmers' Market every Saturday morning with a huge chest freezer in the back of a pickup truck. As soon as they park, they plug the freezer into an electrical outlet. They call their freezer the "Treasure Chest" and have handwritten signs all over their stall that promote the qualities of their beef, pork, chicken, turkey and rabbit. The essence of their farmers' marketing campaign is:

"Clean, chemical-free, local meat straight from our family farm to you."

For years, they had been told by various government officials first one reason and then another why they just couldn't do this sort of thing. None of the officials, however, could agree on the specific regulations that prevented it. More often than not, the Salatins were told "You need to talk to so and so." Then so and so would say "Well, you really need to go through that other office." It was a real bureaucratic mess that just got more and more tangled as the Salatins fought for the privilege to direct market their own meat products.

The end result was that, in this particular jurisdiction, poultry was *exempt* from inspection if the farm produced no more than 20,000 birds annually. Rabbit was also exempt from inspection. Beef and pork are required to be inspected, nevertheless, so the Salatins now haul their cattle and hogs to the nearest USDA inspected processing facility, which is forty miles away. There, the beef and pork is inspected, processed, cut, and wrapped into consumer size packages with clear and precise labeling. Everything is by the book. The Salatins then pick up the end product, store it in chest freezers, and proceed to direct market their own meat products at the farmers' market every Saturday morning. It's a lot of work and an added expense in production, but with the prices that customers are willing to pay for clean, chemical-free, local meat products, it is very worthwhile.

The most popular meat products are: beef steaks and roasts, broilers, pork tenderloin, ground beef, pork sausage, whole turkeys, and rabbit. Another revelation - *the most expensive cuts of meat seem to sell the quickest.*

I'm not going to tell you that retailing your meat products at the farmers' market is easy. It certainly is not. But you should research and find out exactly what you can and cannot sell. Don't be put off by the first government bureaucrat you talk to. Keep trying. Supplying your customers with your own meat products

will be an excellent addition to your market product line.

Specialty Items

Other items that you may consider offering to customers via your dynamic farmers' market operation are:

- **Cut flowers** - Always a favorite at any farmers' market, cut flowers offer several opportunities. Statistics show that flower sales are increasing yearly. Strive to learn how to arrange flowers skillfully, choose varieties that are popular and easy to grow, and how to harvest your flowers so that they will last. Lynn Byczynski's book *The Flower Farmer, An Organic Grower's Guide to Raising and Selling Cut Flowers* is a <u>must have</u> if you are serious about cut flower production.

- **Cider** - Apple cider is a specialty product that requires specialized equipment. If you have your own orchard, and once you are set up with the proper equipment and bottling supplies, the cider business can be very lucrative.

- **Honey** - Almost every market has at least one beekeeper who retails honey. If you are a vegetable or fruit producer, the addition of a dozen hives of honeybees to your farm is almost like adding a business partner. The bees can make the difference not only in a successful harvest by pollinating the crops, but the resulting honey can add considerably more dollars to your bottom line. With farmers' market customers usually more health conscious than average, you should stress the use of honey as an alternative sweetener to highly processed white sugar. Believe me - honey sells!

▸ **Plants-** Many growers augment their income by offering greenhouse products to their farmers' market customers. You might elect to specialize in vegetable plants, bedding plants, herbs, hanging baskets, house plants, or any other of a myriad of choices. Be aware that a greenhouse operation requires skillful and constant management. The initial investment in greenhouse equipment is substantial; however, it is possible, with good management, to gain a *profound return on that investment*. If you dearly love plants and growing things, this will be a natural choice for you.

Summary

Be creative in making additions to your product line at the farmers' market. I've witnessed quite a few imaginative farmers adding to their income by offering such items as gourds, grapevine wreaths, dried bittersweet vines and even pine cones! Consider blown goose eggs, bagged compost, Christmas trees, painted pumpkins, or Indian corn. I've even seen one farmer who sold not only sweet corn in August, but cut the corn stalks in October and sold bundles of them at the market for use as a seasonal lawn decoration. He got $6.00 a dozen for the corn, and then $4.00 a dozen for the stalks! Look around *your farm* and think about the possibilities.

Why These Farm-Fresh Eggs are Superior

prepared especially for our friends at the farmers' market

- Our motto is "Eggs laid by hens that lead a nice quiet life," and that is the principle of how we manage our flock.

- These eggs are produced on a family farm that places an emphasis on nutrition and quality of life. We believe that the quality of an egg is determined by three things: the hen's health, the hen's environment, and the hen's diet.

Our hens *are not*:
- ► fed medicated feed.
- ► crowded together in stainless steel cages.
- ► hyped-up egg machines bred exclusively for output.

Our hens *are*:
- ► fed a chemical-free diet, including fresh greens and surplus garden produce.
- ► allowed loose housing. They can stroll around and exercise, receiving natural sunshine and fresh air.
- ► primarily the heirloom breeds such as Rhode Island Reds and Barred Plymouth Rocks.
- ► ranged on pasture eight months a year and sheltered in a greenhouse during winter.
- ► cared for.

- With the green material found in a "salad bar" type of diet, our hens lay eggs which are up to three times higher in Omega-3 and Omega-6 oils. These are the oils which are found in fish and marine life and have been shown to *reduce* cholesterol. Our feed contains no animal proteins (meat by-products). We also supplement the feed with expensive Icelandic kelp and natural oyster shell. Vitamin and mineral content is significantly higher with this premium pastured diet -- especially the B vitamins which are found in green, leafy vegetables.

- We also believe that **yolk color** is an indicator of health and nutritional level.
Considering color, think about *this* for a moment:

Good	Not as Good as it Could be
whole wheat bread	white bread
broccoli (dark green)	iceberg lettuce (pale & watery)
bright rosy cheeks of a child	the pale face of a sick child
bright orange, farm-fresh egg yolks	pale, commercial egg yolks

We invite you to compare our eggs with any others. Savor the taste - then you'll know <u>why</u> they are superior.

Jeff & Sheila Ishee
Bittersweet Farmstead
(540) 886-8477

Chapter 8

Face to Face With Mr. and Mrs. Customer

Y ou can plant it, and you can grow it. You can harvest it, and you can bring it to the market. You can display it, and you can price it . . . but it's all for naught if you can't *sell* it.

The psychology of sales is a complex one. Of course, we all know natural born sales people. Then there are those that, well, couldn't sell a bucket of water to a fireman. In our modern technological society, we all experience sales situations in one way or another every single day.

Is it possible that you can be a competent "farmers' marketer" without being a competent salesperson? In a word -- No.

Even with a flawless farm product at the perfect market, and with a flawless display on a perfect spring day, it's the "one-on-one" contact that produces the actual cash in hand.

***People centered selling* is the most valuable tool for any business, including farmers' marketers.**

Maybe you've heard the old saying about the three types of sales people, which include:

► **ORDER TAKERS** - The name says it all.

► **CHARLATANS** - They use the fast-talking, hard selling "in-your-face" type of sales technique. Many commonly utter phrases such as "This is your last chance," or "You're going to regret it if you don't buy this today."

► **PERSUADERS** - Those who learn people's needs and then attempt to demonstrate how his product or service will satisfy those needs.

Order Takers

It has been estimated that 90% of all sales personnel are "order takers." You know them. They are the people in a company who stand behind a counter all day. They patiently wait for a customer to walk up with an item, and then, with no charisma at all, register the sale and offer the obligatory "Have a nice day sir."

Haven't you seen it over and over at the market? A lady may be closely examining the different varieties of vegetables on a farmer's table, and (you can almost see it on her face) suddenly she has a *question* about this variety of tomato and looks up to ask the farmer if this is a good salad tomato or a good slicing tomato. But the farmer is sitting on the tailgate of his truck, intent on finishing the sports section of yesterday's newspaper. Eventually, he looks up, squints, and sees the customer standing there with a tomato in her hand and a quizzical expression on her face. He finally asks with a raspy, monotone voice "Can I help you?" She looks off in another direction and says "No. That's OK." She then wanders over to another stall, and the farmer is genuinely relieved. Now he can finish reading the story about Rusty Wallace winning the pole at Talledega. He doesn't really care if he lost one little old sale. "Somebody else will probably want those tomatoes anyway" he figures.

This is a classic example of an "order taker." The farmer is there at the market as if he were just putting in time. This is not *people centered selling*, but rather *self centered selling*. Of course, the farmer will make a few sales each market day, but he is nowhere close to living up to his potential.

Charlatans

After the 90% of sales personnel who are "order takers" comes another 5% who some call "charlatans." I'd wager that most people have certain images of sales occupations that use "charlatan" techniques. Telephone solicitors, insurance representatives and used car sales people are frequently (and sometimes *wrongfully*) identified as belonging to this group. The "charlatan" commonly uses fear as his primary sales technique.

You might be asked "What if your husband, *God forbid*, were killed in a tragic accident tomorrow. Are you prepared to financially provide for the children until they are grown? Are *you* really ready for that if it were to happen *tomorrow*?"

Or you might hear "It's the last one on the lot. In fact, I don't think there is another four-door, forest-green, automatic Accord in this entire state. Two other customers are looking at it, and if I were *you*, I'd sit down with the loan manager right now and wrap this deal up."

Or, how about "If you continue with that other long-distance company, you are just throwing your money down a bottomless pit, and you know that once it's gone . . . it's gone forever. I'm sure you work very hard for your money, don't you! But if you accept our offer to switch over to XYZ Long Distance *today*, you can sleep well *tonight* knowing that you've made the right decision for *your family*."

While all of these phrases may be absolutely true, they place pressure on the customer. And customers under pressure are *not* repeat customers.

Salesmanship requires, above all, the spirit of optimism.

Anonymous

You've seen "charlatans" at the market also I'll bet. As an elderly couple strolls leisurely along the market commons, a farmer jumps out from behind his stall almost *waving* a head of broccoli. He approaches the couple directly and exclaims "I can tell by looking at your shopping basket that you've already gotten several things here at the market today, but I don't see any broccoli in there." Without waiting for a response, he just eases into the "charlatan mode" and skillfully uses the fear technique by saying (rather loudly) "Did you know that doctors recommend five servings of vegetables per day, and that at least two of these should be cruciferous vegetables? And did you know that broccoli and cauliflower are the only two crucifers available at the market this week? You know, broccoli has been proven to (raising his voice) *prevent cancer*. Has anybody in *your* family ever gotten *cancer*? Now, this is the last bunch of broccoli I've got left. Of course, Farmer Brown over there has some, but he wants (even louder) *two dollars a bunch*. Since this is my last bunch of broccoli, and since you need it *today*, I'll let you have this one for only $1.75. Well, how about it?"

If this isn't playing on fear, and putting a customer under direct pressure, then I don't know what is. Regretfully, this technique works on some people. But I think that more often than not, the customer purchases the product just to get the salesman out of his face. If you think that this is going to bring *repeat business*, think again!

"Every great salesman has profoundly at heart the interest of his customer, and no business can develop except as it promotes the interest of those who use its goods or its services"

Charles Schwab

Persuaders

The remaining 5% of sales people, and consistently the *top* sellers in any organization, are the "persuaders." I would imagine you know one personally. Think for a moment. Yes, that's the one! Persuaders are the individuals that always seems to project a positive attitude. They are thoughtful, courteous, and show a personal interest in whatever *you are doing*, no matter if it relates to their business or not. Some persuaders come by their talents naturally, having been raised by like-minded parents who instilled the Golden Rule in them years ago. Other persuaders obtain their skills by cultivating good personal habits and by studying the characteristics of successful people around them. They work at it. It's just like a farmer cultivating his soil for future crops. He may not have been "blessed with good soil", but it is still obtainable through hard work.

Of course, becoming a "persuader" is *not* like learning to drive a car, or even learning to line dance. It's not a technique at all, but an art. But like an artist, you can always learn new methods.

Physical Appearance

Your own attitude towards yourself and your farm's image are conveyed to potential market customers in a variety of subtle ways.

> ► **Eye contact** - Exhaustive studies show that people who make eye contact, and then hold it, are people you can trust. When the eyes dart away after initial contact, human nature seems to throw up a big sign

saying "beware." When you are at the farmers market, always, but *always* make eye contact with a potential customer and *hold* it! This, of course, does not mean staring or glaring. But keeping eye contact with people tells them one thing . . . that you are interested in *them*. You can practice this by looking at yourself in the mirror.

▸ **Body language** - How many fellow marketers do you know who bring a lawn chair with them to the market? Not only a lawn chair, but the daily paper, a portable TV, and a paperback novel? As soon as the market display is set up and the produce is unloaded, this marketer sits down (facing *away* from the customer area), puts his feet up on an apple crate, lights a cigarette, and retreats into his own little world. You or I, if we were customers, *wouldn't dare interrupt* this person to buy a head of lettuce. We know that the lush green lettuce on the second shelf of the farmer's display seems really fresh and attractive, but most folks would walk right on by. It's simply not worth disturbing this person and entering his *comfort zone* just to buy a head of lettuce.

Negative body language includes:
- Arms folded across the chest
- Sitting while the other person is standing
- A frown, or a facial expression of boredom
- Not directly facing the person with whom you are engaged in conversation
- Hands in the pockets

Positive body language includes:
- Nodding your head as the other person speaks
- Hands clasped together either in front of, or, preferably in back of the waist
- A genuine smile
- Shoulders squared towards the person you are talking to

▸ **Clothing** - One thing I like about being a market gardener and a "farmers' marketer" is that I can wear really comfortable clothes. I'll take a pair of Wranglers and a sweatshirt any day over a three piece suit. What you wear to the farmers' market *does* make a difference. The physical image your clothing projects should be that of health, color, vitality, and cleanliness. Clean jeans, a cotton shirt, and a baseball cap go a long ways towards meeting these perceptions. This "uniform" is suitable for both men and women. Overalls are fine, because they have the image of "farmer" written all over them. It's OK to wear something distinctive such as red suspenders because people will remember the colorful and unique attire; however, stay away from faddish clothes at the farmers' market. People come to the market for customary purposes, not to see an outlandish fashion show. Wear your old boots or garden shoes, and not brand new, slick, white, basketball high tops. Somehow, Reebok and Nike do not symbolize fresh and local farmproducts. Old leather boots do.

A persuasive and dynamic farmers' marketer is one who uses all the tools in his toolbox when it comes to making a sale. He will remember his customer's name because he knows that people love hearing the sound of their own name in public. He will wear conservative, yet casual, clothes because he understands that potential customers might not willingly identify with a "hayseed redneck." But when face to face, they *can* feel comfortable with a "country gentleman."

"At one of the markets not long ago I asked a twelve-year-old friend to mind the stand while his father and I settled down on the tailgate to catch up with each other's news. There was a lull and I knew the boy was capable of dealing with the odd half a pound of garlic that might now and then be handed to him to weigh and price. But he objected. In a slightly panicked voice, he said 'I've never sold anything before.'

At his age, and even much later, it would have been my protest as well. In our society the young are trained early to buy. Selling comes later."

Stanley Crawford
A Garlic Testament

Chapter 9

Visual Displays That Draw A Crowd

Your stall at the farmers' market is your visual calling card. It is the *display* of goods that the customer sees first, and this is something that most patrons will unconsciously compare with other stalls at the market.

Let's be candid here. Most of us do it without even noticing it. We stroll down the mall looking at storefront displays, looking for what suits our mood, what suits our tastes, and what suits our wallet. When we see something that strikes our fancy, we walk into the store. Now we are at that place the retail industry terms the "point-of-purchase." Of course, the savvy merchant has a *different* display at the point-of-purchase. Normally it is a display with more *details* about the product and more *suggestions* as to why we *need* that particular item. Marketing is a fun little game, don't you think!

When I chat with fellow farmers at the market and talk about display, some of them

Displaying your goods can be:
    √ **Versatile.** You can mold your display to fit the season, your product, and your farm's image.
    √ **Economical.** We are not talking about big, fancy, department store display techniques here. We're talking about basic materials, the space that you utilize, and the product you want to emphasize.
    √ **Productive.** Your farmers' market display is readily visible to anyone who passes by. There is no time-lag between when a potential buyer sees your stall and when he can react to it.

say "Humph. I don't need a fancy display. I can just throw my stuff on the sidewalk and it will sell itself." Indeed, the word "display" conjures up images of elaborate, expensive arrangements in fancy, big-city stores . . . but it doesn't have to. Displaying your goods at the farmers' market is a simple tool that you can use to help increase your gross sales, *and* your customer loyalty.

At most markets, there is an area that the customer strolls through that I'll call the "commons." The commons may be a walkway, a parking lot, a covered patio, or a street. The customer walks the commons of the farmers' market, looking for something appealing, something attractive. When he stops, and takes a step closer to a farmer's stall, he has seen the "window display," and is now approaching the point-of-purchase. The customer doesn't see the *design* of the display. He sees *what he wants to purchase*. Your job, as a dynamic farmers' marketer, is simply to steer him to what he wants to purchase.

Emotional Appeal

The first thing a farmer's display should do is give shoppers a reason to pause and look. This means catching their attention with something interesting or appealing. And you've only got about three seconds to do it! Catching the customer's attention can be done quite effectively by incorporating an *emotional* appeal.

Read the last sentence of the previous paragraph again!

We did this at our own stall one year by placing a small handwritten sign next to our fresh brown eggs. The sign was nothing but a little 4 inch by 6 inch beige card with red

magic marker declaring:

"EGGS LAID BY HENS THAT LEAD A NICE QUIET LIFE."

It never failed to cause a smile on potential customers' faces when they saw that little sign. This verbal suggestion told the potential customer a lot about the size, the spirit, and the overall attitude of our farmstead. It implied "Family Farm," "Healthy Livestock," "People that Care." It appealed to emotion and told the customer that there was a quality product here. We found that more times than not, people would buy into this ethic . . . and smile as they paid for a dozen eggs! (Of course our hens *do* lead a nice quiet life. They thrive on pasture in spring, summer and fall, and they have their own greenhouse in winter. Truth in advertising is essential.)

Colors in your farmers' market display can assist you in appealing to emotion also.

Consider the following:

-**Yellow** feels warm and light. It works well in mid-summer.
- **Green** feels natural and fresh. It works great in springtime.
- **Earth tones** feel natural, and border on sophistication when you target young people.

Image

Your display should tell your customer about your farm. It should project a certain image. It is this image that compels the shopper to initially patronize your stall. The quality of your product, the price, and your ability to be consistent will keep the customer loyal, but don't ever forget that *display* is what "gets them in the door."

Our first year at the local farmers' market, we made up a tri-fold poster board with photographs of the farm. We included pictures of us tilling the soil, washing pumpkins, the kids playing in the barn loft, and the quaint gravel lane leading up to our farmstead. We included captions below the pictures such as:

- *Summer kitchen, circa 1900, soon to be renovated for our on-farm bakery*
- *Little North Mountain as seen from the pasture*
- *Bittersweet Lane, near Middlebrook, Virginia*
- *Cold crisp morning from the "sun porch"*
- *Three good kids, all "out standing in their field"*
- *Garden preparation last fall*
- *Portion of old orchard, which includes apples, cherries, pears and peaches*

Also on the poster board was an announcement proclaiming :

"Coming this summer and fall: raspberries, blackberries, tomatoes, squash, apples, pie pumpkins, and, of course, more of Sheila's delicious baked goods."

Essentially, we took our customers on a visual tour of the farm and let them know who we were. We introduced ourselves at the market with this informal display and immediately people seemed to comment positively about our farm operation and the family emphasis, even though they didn't know us personally. Older folks would say something like "Oh, I can remember growing up on a farm and playing in the barn loft. We stored our apples up there. Say, how is the apple crop coming along this year?"

Young couples would look at the pictures and say "Wow! That's neat. Raspberries are coming in soon, huh? Are they hard to grow?"

We intentionally portrayed an *image* which conveyed three things:

▸ *family oriented*
▸ *locally grown and fresh*
▸ *people having fun.*

As a dynamic farmers' marketer, you should ask yourself what image *you* want to project at the market. It may be different from ours. Depending on the market, *upscale and modern* may be just what the customer wants.

Let's say that you sell at the market in midtown corporate Philadelphia, where the majority of your customers are thirty-something, trendy, and wear a business suit 12 hours per day. In this situation, modern display fixtures and fashionable clothes (*L.L. Bean* boots and a *Land's End* jacket?) may be appropriate. If you appeared at this market wearing bib overalls and a straw hat, sitting on a rusty tailgate chewing a big wad of tobacco, you might have more people gawking and snickering than actually shopping.

Visual impressions are critical in creating an emotional appeal with your customer. Keep in mind that sometimes it might *actually be* fresh, healthy food, and nothing else, that the meticulous customer is after, and not a two minute venture to backwoods America. It all depends on your market situation.

An emotional appeal normally causes people to stop and take a second look. Your market display is simply presenting information to an observer. What kind of display you use, or design, is up to you. When potential customers see a design and it has a *mood*, they respond to it. It arouses their curiosity.

How Corporate America Appeals to Emotion

Consider a big city department store window (at night) featuring a striking ravenhaired mannequin wearing a lowcut red dress and high heels. An Italian leather purse dangles from one shoulder. The mannequin's chin is tilted slightly upward. There is no other fixture in the window. A single dim spotlight emphasizes the almost surreal figure. This mannequin seems to whisper "I'm going to have fun tonight. Are *you*?" If a young woman walking by is receptive to this obvious mood, then the display arouses a sense of adventure and she is more likely to go in and buy the dress (or one like it). This is a good example of marketing for emotional appeal.

Farmers' Market Adaptation

So you ask "How do I use this technique to help me move my sweet potatoes?"

The answer is that you have to appeal to the customer's emotions concerning good health. You might make up a small handwritten sign that simply states "SWEET POTATOES PROMOTE GOOD HEALTH!" This is a direct appeal to the customer's emotions. A lot of potential customers will let curiosity get the best of them, and they will ask "*How* do sweet potatoes promote good health?" Of course, then you tell them that recent research by the Center for Science in the Public Interest established the sweet potato as *the* most nutritious vegetable, getting a score more than twice the next highest vegetable. If they seem interested in this, go on to tell them that sweet potatoes contain four times the U.S. Recommended Daily Allowance of beta-carotene, twice the RDA of Vitamin A, and 30% of the RDA for Vitamin C. Then you hand them a 3 x 5 card with a simple recipe for sweet potatoes and say "Now. This recipe is one of my favorites. It's low fat, delicious, takes only 10 minutes of preparation time, and

---

"People don't buy our services, products, or ideas. They buy how they imagine using them will make them feel."

Spencer Johnson, M.D.
The One Minute $ales Person

---

calls for 4 pounds of sweet potatoes. You know, your friends that are health-conscious like you might enjoy this recipe also, so feel free to share it with them. Here- let me get a bag for you. Do you think four pounds will be enough to last until *next week's market*?"

Not only does this use the appeal to emotion, but it uses the power of suggestion.

Simple hand lettered signs such as this can be very effective. This sign tells what the item is, describes it distinctly, and tells the customer about the most important aspect -- taste.

Countless times at the market, my wife would be sharing a recipe with a customer, and, within a minute, it seemed like a crowd would gather, all trying to peek over her shoulder at the same time.

Display Fundamentals

Physical characteristics of your farmers' market display are vital to the success of your effort. *All* displays should have the following common features:

❶ - A distinct sign with your name or the name of your farm.
❷ - All items should be clearly visible and easy to reach.
❸ - *Everything* must have the price clearly marked.

A well-stocked display is always an attention getter. As the volume of your inventory goes down during the market day, keep changing to smaller and smaller containers. You must provide an illusion of bounty. Start with potatoes in a bushel basket, then go to half-bushel, peck, and so forth. Nobody wants to get the last of anything, and when you have four potatoes in the bottom of a bushel basket, it's not appealing. Customers may think "Hmmph. Everybody else has already picked through that basket and gotten the best. All that remains now are those leftovers, and I'm not going to buy *those!*"

Tip your baskets and containers slightly so that it appears produce is so plentiful that it is spilling out. This illusion of plenty creates a "bigger choice" for the customer, which is

69

Good Morning,
Welcome to the Market

Our Greenhouses are located
on Buttermilk Spring Road

We invite you out to inspect
our Plants and Facilities

Croi-B-Crest
Farm

**Computer generated signs can be effective also,
especially elegant and conservative signs like this.**

what all people want . . . more choices.

The Power of Signs

Signs are a must at your farmers' market stand. One or two signs should be legible from at least fifty feet away. You'll probably want your farm's name on one of these signs. The other might be a "special of the week", particularly if you are the first at the market with something just coming into season (sweet corn, fresh tomatoes, etc.).

Small handwritten signs are one of the most useful things you can use. Beige or earthy green colored card stock works well, but white is fine also. I've found that 4" x 6" is about the best size. The beauty of these little handwritten signs is that you can get a great idea, write it on the sign, and then display it next to your produce . . . all in less than a minute. Every

farmer should have a supply of blank card stock on hand as part of the marketing effort.

People never want to appear ignorant by pointing to a vegetable and having to ask "What is that?" It's better to have a small sign next to your eggplant that states **"Italian eggplant - grown organically - makes a wonderful casserole - ask for free recipe - $1.50 per pound."** You might even add (in small print) "Three pounds is enough for a nice size casserole." Hand-lettered signs and labels are important in the success of your farm stand at the market.

"But what about computer generated labels?" you ask. If you are dealing in a commodity that *must* be labeled (by government requirement), then certainly you must have a professional label that meets all the criteria set forth. Examples of this may be

70

egg cartons, meat packaging, etc. If you are selling a high *quantity* item that uses a lot of labels, you probably would want to use computer generated labels also.

When my wife makes several hundred baked goods for the market, there is no way I would give up my computer labels (30 per sheet in about 15 seconds) and hand label each item. The only thing you must remember is to use legible fonts that can be read effortlessly. Do not use script or exotic fonts when generating your labels. Good fonts are **Times New Roman**, **Arial**, or other standard types.

But I also believe that when an item *can* be hand labeled, it *should* be. If there is a slow period at the market, this is a great time killer. It makes you look busy, and builds an inventory of labels for future use.

Simple, Yet Effective, Fixtures & Decorations

Use a lot of natural farm "fixtures" in your display. Suggestions include:

- Keep merchandise off the ground by using bales of straw as a shelf.
- Drape burlap over cardboard boxes.
- Lay an old (but clean!) board between two upside-down bushel baskets and use this as a shelf. Just behind this, lay another old board between two taller bean baskets. This gives a "stair-step" appearance, and makes it mush easier for the customer to see your produce. All market displays should have at least **three vertical levels** in the display.
- Wooden apple crates are a marvelous tool to use for displaying a wide variety of fruits and vegetables.

- Burlap makes a remarkable tablecloth that suggests the feeling of "farm fresh" via its color, scent, and texture.

You should use groups of products in your display that are related to each other in the customer's *end-use viewpoint*. Place radishes next to lettuce, for instance. Or fresh peas alongside small new potatoes. Another example would be to put fresh strawberries next to shortcake or poundcake. This grouping effect has the power of suggestion, and is a very effective way of displaying your farm products.

Be Creative

Creative methods to enliven your market display might include the use of cut flowers in a galvanized feed bucket. You might use picnic baskets with gingham napkins for baked goods. Also, brightly colored produce is truly appealing when piled high in little red wagons.

One of the most ingenious displays I've ever heard of utilized fresh brown eggs arranged loosely in a bed of chopped straw. The straw was atop a tray of ice cubes (covered with plastic film). This held the eggs at proper temperature. A hand-lettered sign announced:

"PICK YOUR OWN - 20 CENTS EACH, OR 2 DOLLARS PER DOZEN."

Empty egg cartons were adjacent to the display. Customers had the sensation of actually gathering fresh eggs from the "nest," and *all the farmer had to do was keep the nest full.*

The beauty of farmers' marketing is that nobody else has control over your *imagination*. You are on your own! After you meet the physical needs for a display, let your imagination run free. Every morning you set up, step back a few feet and look at it from the customer's perspective. Does it appeal to emotional needs? Is it colorful? Is it convenient? Does it have the power of suggestion? Does it portray the image that you want?

If so, step back around to the other side and get ready for a successful and *dynamic* market day!

Samples of Good Signs

**If you specialize in a certain commodity,
let the customers know with sizable signs.**

Be unique. People love it.

Very powerful sign when placed within a five-block area of the market. This sign causes drivers to impulsively think "Well. I'm this close, so I might as well stop in and see what's at the market today." Within 60 seconds, they are at the market.

Chalk boards and dry erase boards are wonderful tools for your market business.

Appendix A

Stories from the Frontline

NOTE: All of the following information was submitted freely to the author of this book, or obtained from public domain documents on the World Wide Web. Credit is given where the originator is known. Most of this material is wonderfully written, so pay attention - these are stories from the frontline!

Some time ago, there was an article in the Atlanta Journal-Constitution about closing down the City Market. I don't know whether they did or not. If it's still there, this is a Farmer's Market slap dab in the middle of downtown Atlanta -- about 3 blocks from the State Capitol. Twenty years ago, I worked at the large hospital next door and would wander over there at lunchtime for wonderful fried oyster sandwiches in season and the freshest and cheapest fish in town. It was wonderful. I hope you have also talked to the people who buy from the Farmer's Market and then take the produce to the side of the road in outlying areas and sell them. My sister-in-law's father would go to the Atlanta Farmer's Market near the airport and buy produce, then take it to the Athens Farmer's Market 50 miles away to sell it. Years later, after he had died and her mother remarried, her stepfather filled his pickup with whatever was in season and had

his "space" on the side of the road where he sold it. Both the people selling and the customers were very territorial. If he was not there, his customers got upset and wondered if anything was wrong. They counted on him for tomatoes in summer and apples in fall because they knew him and knew his produce was fresh, high quality and a good value.

You may have made a mistake in coming to the (Internet site) "GardenWeb." The people there are so knowledgeable and helpful and so full of ideas that you may discover your book is not nearly finished after all. Good luck. When your book is published, please post it for all of us.

Anonymous e-mail
received by the author

I've been selling at our local Farmer's Market off and on for almost 10 years. I don't know about other states, but did you know it's mandated by law that a place be provided in each town in Arkansas for a Farmer's Market? Not all of them have one, but a lot do!

I found the Market when I was broke, had a big garden, and found out it was free! Can't get better than that. In our little town it's on the courthouse square every Wednesday and

Saturday morning. All you have to do is pick & pack your veggies (or fruits, or jams & jellies, or eggs, or handmade crafts), plump yourself down in an empty spot and prepare to smile and sell. As usual, gardeners are among the nicest people in the world, and the customers are the next nicest.

My best sellers are always green beans & tomatoes. It's sometimes hard to compete with the big growers of tomatoes, who sell to farmers to sell locally. I grow bush beans, pole beans and Italian beans.

If I had a strawberry patch or a cornfield, that would be good, too. Watermelons and cantaloupe are big sellers, but they bring most of them from further down south in Arkansas. I sell all the lettuce, fresh peas, sugar snaps and snow peas I can grow in the early spring, because they don't last long. Also all the farm fresh eggs I can spare.

If there's anything else I can help with - let me know. Gotta run. Good luck! It's a fascinating subject.

Anonymous e-mail
received by the author

Q - *How do you price your product?*
A- percentage of cost to produce.
Q - *What are your top 3 best-selling items? Why?*
A- Emu jerky
   Emu oil
   Emu Salami
Because the meats are awesome tasting, and different, low fat red meat, and the oil is the natural healer of the 21st Century.
Q - *How did you get started in farmers' markets?*
A- Living in the country, it made sense for us to market in the manner. Especially if your products are out of the ordinary, and you don't

mass market.
Q - *Describe your market (location, # of vendors, season, etc.)*
A - Season: year-round by computer (holiday season is good also).
Q - *The most important thing I've learned about farmers' marketing is . . .*
A - Be flexible, be friendly, being aggressive helps (not too much) but don't take a back seat, or you won't sell anything!

Margaret and Steve Pounder
Spring Creek Ranch
Kooskia, Idaho

Great Quotes Relating To Farmers' Markets *(provided by OpenAir-Market Net)*

1. A good start for making a town's acquaintance is to visit its morning markets, and I do not mean market halls but the temporary kind that consists of collapsible stalls put up for a few hours in a street or a square. Disagreeable as shopping in inclement weather may be to people addicted to the indoors, open-air markets are more than a match for hermetic stores. There is nothing like fresh air and daylight; the knowledgeable shopper prefers the least tidy street market to the air-conditioned morgue of a supermarket.
- *Bernard Rudofsky (Streets for People: A Primer for Americans. New York: Doubleday, 1969. p. 201).*

2. In countries with a respectable architectural past, daily markets are usually located in a towns most sumptuous section. They are the hub of human activity, breathing none of the funeral atmosphere of latter-day civic centers.
- *Bernard Rudofsky (Streets for People: A Primer for Americans. New York: Doubleday, 1969. p. 207).*

3. The Open-Air Market ... It is in a place where it is like summer all the year around.
- *Jane Miller (To Market We Go - 1st Grade Reader. Boston:Houghton Mifflin, 1935. p. 41).*

4. First, the economic success of markets rest primarily on their location and their environmental quality. That quality is defined by 'vibrancy', colour, hygiene, and the spatial experiences of users. It is not necessarily related to the level or type of infrastructure. Indeed, some of the least successful markets found in Asia occur in extremely expensive, highly serviced, formal market buildings.
- *David Dewar and Vanessa Watson (Urban Markets: Developing Informal Retailing. London: Routledge. 1990. p.54).*

5. While wealthier people generally have access to private transport and can easily satisfy their shopping needs, poorer people are usually confined in their consumer behavior to their local areas. For poorer people, access (by foot) to the range of basic and often cheaper products provided by a market can significantly improve the quality of their lives.
- *David Dewar and Vanessa Watson (Urban Markets: Developing Informal Retailing. London: Routledge. 1990. p. 27).*

6. For many people market shopping is not an opportunity for bargains but the realization of a romanticized view of one facet of their cultural past. The simple respite from cellophane is enheartening.
- *Jack Pasternak (The Kitchener Market Fight. Toronto: Samuel Stevens, Hakkert & Co..1975. p. 26).*

7. They could be gorged with any combination of apple butter, schmearkase, schnitz, colby cheese, lettuce, plums, eggplant, granola, shoofly pie, goose liver, halvah, radishes, poppy seed coffee cake, kochkase und kimmel, schwadamahga sausage, bread, or maple sugar fragments. Not only were the baskets replete, but each week the sensory experience of the sights, smells, and sounds of the market would stimulate afresh the "unurbanized" segments of one's consciousness.
- *Jack Pasternak (The Kitchener Market Fight. Toronto: Samuel Stevens, Hakkert & Co..1975. p. 27).*

8. The social ecology consisted of the independence of many people in the market area. Old people depended on farmers for some of their food. Farmers relied upon shoppers to help retain an agricultural way of life. Children learned, through exposure to different kinds of people at the market, useful lessons about life and society.
- *Philip Langdon, et.al. (Urban Excellence. NY: Van Nostrand Reinhold. 1990. p. 23).*

9. At Pike Place, people take pleasure in a relatively humble architecture that connects them to the past and provides plenty of opportunity for social interaction. ... A market can offer the prospect of racial, ethnic, and economic integration-better, probably, than any other part of a city. Cities can benefit from such strong, humane, functional focal points.
- *Philip Langdon, et.al. (Urban Excellence. NY: Van Nostrand Reinhold. 1990. p. 62).*

10. It is good, for instance, that the farmers and shops at Pike Place can serve multiple constituencies -- low-income city residents, gourmets, and restaurants among them. This enhances economic opportunities for farmers and independent local businesses. This is a

worthy alternative to "festival markets" which cater to fewer needs and a narrower clientele and therefore offer less long-term satisfaction.
- *Philip Langdon, et.al. (Urban Excellence. NY: Van Nostrand Reinhold. 1990. p. 62).*

11. Open-air markets are essentially expressions of economic democracy where there is no latent fear of any single vendor dominating the market and forcing others out of business. Merchandise quality, service, price, etc. are the main factors determining survival and success, which is as it should be.
- *Richard Leader (rleader@gv.net) in an e-mail to OPENAIR-MARKET NET, December 1995.*

12. I have always thought that open-air markets serve a function similar to that of the Internet - which is to give leverage to small producers.
- *Sheila J. Siden (sjsmba@netcom.com) in an e-mail to OPENAIR-MARKET NET, January 6, 1995.*

13. Potomac Mills and many other discount outlet centers have become so popular with foreign and American tourists that they are now bigger attractions than the Liberty Bell, the Jefferson Memorial, the Alamo, and many other traditional tourist treasures... many just want to shop, planning vacations solely around that passion. The Travel Industry Association of America recently said shopping was the most popular activity of vacationing Americans last year, and Mr. McMahon has an inkling of the reasons -- the only time many people have to shop, he said, is when they are on vacation.
- *Edwin McDowell, New York Times, 5/26/96, p. 1 and 17.*

14. Shoppers needn't feel guilty that they are not taking advantage of the cultural aspects of the city by spending their time shopping. Don't let anyone tell you that the two don't mix. In the bustling markets of Bangkok, shopping is definitely a cultural experience.
- *Johan Bunger, of Sweden, from his 'Markets in Thailand Page"*

15. A Farmers' Market is a delightful counterpoint to modern life, a little patch of green in an asphalt city, an oasis of sight and touch and smell in a climate-controlled, vacuum-sealed world. Having been eclipsed by the glamour of the supermarket some 50 years ago, farmers' markets are flourishing again. Direct contact is the lure of the farmers' market -- direct contact with the growers, with the produce and, if one is lucky, with one's appetite.
- *Molly O'Neill, "Market Value", New York Times Magazine, June 9, 1996, p. 151.*

16. Everybody loves them. When the market is on the street, crime is not. It kicks the drug dealers out.
- *Caroline Shoenberger, Chicago Commissioner of Consumer Services. quoted in "Farmers' markets reap fresh success across the nation." USA Today, August 23, 1996*

17. Retail tourism, where shopping becomes the main, if not the entire draw for travelers, is spreading across the globe - and reshaping tourism. The U.S. Travel and Tourism Administration found that 84% of international travelers to the U.S. rate shopping as their favorite travel activity. Domestic tourists rate shopping No. 2, behind eating.
- *Scott McCartney, "In Guam, Shopping*

Sprees Are Replacing Tanning." Wall Street Journal, August 23, 1996, p. B4

18. By law, no more than 4,000 food and 1,700 merchandise vendors may operate on the streets of New York. Yet, the 18,000 currently working as vendors without permits testifies to the opportunity in this occupation. Overly restrictive laws force bootstrap entrepreneurs to operate underground. All bootstrap capitalists need is the opportunity to earn an honest living. That is not too much to ask for in America.
- *William H. Mellor (President of the Institute for Justice) , "No Jobs, No Work." New York Times, August 31, 1996, p.*

19. My feeling is that in terms of microenterprise, economic stability, and, also, overall mental health in so many intangible ways, markets make a tremendous difference.
- *Naomi Holloway (starcrow@olympus.net) in an e-mail to OPENAIR-MARKET NET, November 1996. She is also the author of "A Case for a Mothers' Wage."*

The Vineyard Farmers Market: Past, Present, and Future -- An Idea Takes Shape *(used with permission of author Sharon Young. Excerpt from her book "Abundant Harvest.")*

Richard Erganian grew up surrounded by produce, so it's really not surprising that he decided to open a farmers market in the parking lot of the family-owned Vineyard Shopping Center. His first grower, Jimmy Hirasuna, made a grand total of $9 selling vegetables and early fruit at the Blackstone at Shaw site on the market's opening day, May 3, 1980. By the middle of July, more than twenty growers were gathering every Saturday to offer the bounty of their trees and fields to Fresno-area shoppers hungry for just-picked fruits and vegetables.

Richard's parents, Yervant and Askanoosh Erganian, arrived in Fresno on November 1, 1920 from their home in Bitlis, Armenia. They went into business with a partner who operated a small neighborhood grocery store near the old Emerson School, west of the Santa Fe railroad tracks. The couple worked long hours learning the grocery business and going to school to learn English.

In time they saved enough money to buy a piece of property, which included a larger grocery store and several houses, in the 600 block of O Street. The Erganians settled in one of the houses and rented out the others. Their O.K. Market outlasted all the other stores established in the area over the years, expanding three times before succumbing to redevelopment in 1983.

After construction of the Fresno Convention Center and Highway 41, the remaining neighborhood was razed and the land given free of charge for Pacific Gas and Electric offices. Richard then moved the family home from 644 O Street seven miles north to a ten-acre site on the northwest corner of Blackstone and Shaw Avenues. The Vineyard Farmers Market now occupies the center quarter-acre of the property.

The grocery stores of those early years bore little resemblance to today's supermarkets. In the days before home refrigerators, stores opened early and closed late to meet the needs of families that counted on them for milk and other staples. For the Erganian children--sons Aram, Richard, and Miche and daughter Aznive--the O.K. Market was an extension of their home. The family worked in the store virtually around the clock, bottling wine and olives, making shortening and Armenian specialty foods, and stocking the store with produce bought several times a week from local growers who gathered at 4 a.m. at a wholesale market on G Street next to the Southern Pacific Railroad tracks. They even made deliveries of food and kerosene to customers who lived several miles from the store.

Armenians hungry for a taste of home could satisfy their longings with the Erganians' Basturma and Soujouk (meat snacks), Roejig (a sugary walnut treat; see recipe section), and imported items such as figs and pistachios. It's difficult to imagine now, but California did not begin marketing its own dried fruits until the 1940s--and pistachios were not grown here commercially until 1976.

Richard remembers how much he loved the downtown Fresno farmers market when he was growing up--the color and excitement of the crowds of people buying and selling everything from garden-fresh carrots to live chickens in the area around Courthouse Park. "It was a worthwhile experience for children and parents alike, a real community gathering place," he said. Those early experiences, coupled with an introduction to European farmers' markets on visits to relatives in Paris and an interest in architecture and landscaping, played an important part in his decision to establish a market in Fresno.

Turning the Idea into Reality

Starting a certified farmers market in Fresno wasn't nearly as easy as Richard had imagined. "I felt that since Fresno was always being touted as the agricultural capital of the world we could certainly do a farmers market. What I quickly found out was that over 90 percent of the farmers in Fresno were set up on a commercial production scale to sell to brokers or wholesale houses and not get involved with direct sales themselves."

Richard had plenty of business savvy. With bachelor's and master's degrees in business administration, real estate credentials, and a career with Wells Fargo Bank behind him, he knew the risks involved in this kind of venture. "I had an awakening when I discovered there were very few family farmers around Fresno set up for direct marketing. Most farmers who don't grow cotton grow grapes, which travel by ton truckloads to wineries and produce houses all over the world. Next comes tree fruit that is moved by trucks all over the world. Very few farmers are set up to sell at a local farmers market. So it took a lot of looking, a lot of persuasion, a lot of compromise. The first year I didn't even charge, of course. I sensed that growers were suspicious of me. As I approached growers, they were thinking: `Why is this guy starting a farmers market at Blackstone and Shaw? What is he up to?' There was an eerie sense of distrust. I had already figured out my arithmetic and I knew that I wasn't going to make money at this thing, so I told them: `Try it free for the first year. If you like it, come back.' My remuneration for the first year was an occasional watermelon or a bag of peaches."

The Vineyard Farmers Market's first season was a learning experience. Richard started with a Saturday market, but he invited growers to bring their produce out to Blackstone and Shaw seven days a week. A mailing recruiting area farmers to participate in the market's second season stated: "We are open seven days a week from 12 noon to 4 p.m. Choose one or more days to fit your schedule and bring your produce to Fresno for quick cash sale." Noting that 25,000 cars pass the market location every day, the mailing pointed out that at the market growers could eliminate middlemen, set their own prices, and get some immediate cash.

Saturday was by far the busiest day at the market. In a column published in the July 13, 1980 Fresno Bee, Woody Laughnan described a Saturday at the market, where one could find "just about every summer crop that is grown in the valley." The market was apparently a lively place the day Laughnan visited, complete with street musicians. Growers were selling their produce from the backs of pickup trucks, campers, vans, and travel trailers. At a time when they could make only a limited profit selling to packing houses, many small growers looked to the farmers market as an opportunity to make some much-needed cash. A woman selling farm-fresh eggs explained: "Egg prices are down and we have to make some extra money any way that we can."

The Vineyard Farmers Market's first season ended in October, as the summer fruits and vegetables ran out. On the first Saturday of October, before the market closed for the season, a Fall Harvest Festival celebrated the Valley's bounty. A highlight of the event, which was repeated the following year, was children carrying bunches of purple balloons intended to resemble grape clusters. "My fingers were blistered from tying all those balloons," Richard recalls.

The growers were back at the Vineyard Shopping Center parking lot ready for another year the first Saturday of May 1981. An article in the May 3 Fresno Bee noted the opening of the market, which the Bee said would run on Thursdays from 4 to 8 p.m. and Fridays and Saturdays from 8 a.m. to noon. Twenty or more growers came out that first Saturday, and as many as fifty were expected to participate during the season.

In contrast to the market's opening day a year earlier, a grower could sell $100 worth of

produce if he or she had sufficient variety. Many Central Valley growers were not interested in a Fresno market, however. They could drive to farmers markets in the San Francisco and Los Angeles areas, sell a larger volume of produce, and charge higher prices for the same produce than they could get in Fresno. This is still true today.

The Market Matures

Shoppers who have experienced the Vineyard Farmers Market only in its current grapevine-shaded arbor setting may well wonder how the growers endured those first four summers standing under burning sun on top of scorching asphalt. It wasn't easy.

The produce as well as the growers suffered. The growers had been up since 4 a.m. and put in a full day by noon. When the market closed they would have at least another hour's work returning to the ranch, unloading, and so forth. Many of them still had irrigation and afternoon chores. Customers weren't inclined to linger and chat with growers and friends as they do now. They tended to "hit and run" to get out of the 105-degree heat that felt like 125 degrees. Even though growers rigged up all kinds of shades and umbrellas to keep off the sun, nothing could really dissipate the heat of a Fresno July afternoon.

Like everything about the Vineyard Farmers Market, the idea of putting up an arbor to make the shopping experience more comfortable for growers and shoppers alike was part of an evolutionary process. Richard credits local artist Alyce Stukenbroeker, a faithful market shopper and friend who died in 1994, with giving him the idea of an arbor.

"We spent hours discussing philosophy and exchanging books on health, gardening, and nutrition. One day we were looking at photos of an English garden with wisteria growing over an arched structure," Richard remembers. "Alyce sketched a larger scale version with a pencil. We essentially took that drawing and superimposed it on a set of parking bays."

Richard knew he couldn't charge stall fees high enough to finance a large building. "Even if we could have put up a building, we'd have had to use air conditioning--and the growers couldn't pay for that," he explained. "That was my first step into sustainable architectural design. I had already had experience with grapevines in the landscaping of the Vineyard Shopping Center. I thought what would make an ideal shade structure was taking the idea of that English structure and substituting grapevines."

Grapevines were the ideal choice for two reasons. First, grapes are one of Valley's primary crops. Second, and more important for the arbor, grapevines provide excellent shade. The buds start coming out in mid-March, just as the temperature starts to climb and shade is needed. The leaves get larger as the weather gets warmer, and by June the vines completely shade the area. The leaves begin to drop in the fall, opening up the arbor to let warming sunlight through during cold winter days.

Getting from concept to arbor took some time. A local architect designed a shade structure that was basically a carport covered by a trellis, but somehow it didn't click. That meant another hot summer on the asphalt, but more and more Valley residents were learning about the market and business was good.

As increasing numbers of growers brought their produce to the market, Richard learned something important. "In May or June when new growers asked if I had a space, I would just open up one more parking stall," he said. "I think at one point I had fifty growers out there. I started to look at the situation in the middle of July, and I realized that half of the growers were selling the same thing--peaches, plums, and nectarines. I realized it didn't make sense for either the customers or the growers. If there were twice as many sellers as we needed, each was doing half the volume." From that point on, he concentrated on encouraging growers who were offering something new or different.

The Arbor Emerges

In 1982, while taking a class in city planning at Fresno State, Richard came across a book on pattern language by Christopher Alexander, an architect at the Center for Environmental Structure in Berkeley. "His spirit of design made me realize he had the right heart. I said, 'This guy is writing down what I'm trying to do here at the market.'" He telephoned Alexander, who agreed to take on the project.

Alexander and his structural engineer, Gary Black, interpreted Alyce's sketches and Richard's wishes and labeled them "Les Halles," after the Paris farmers market complex established in 1854 and demolished for redevelopment in 1971. "What Les Halles was to Paris, I hoped the Vineyard Farmers Market--which I call Petite Les Halles--would become to Fresno," Richard said.

Over a period of six months Alexander, Black, and three graduate students developed the design and engineering for the arbor. The following year, they fabricated the redwood elements of the structure in a warehouse in Berkeley. Two graduate students built the concrete block columns on the Vineyard site, and then the project sat idle for a couple of months while redwood beams were being constructed.

"I remember a lot of comments, because we were still having the farmers market in the parking lot," said Richard. "People thought I was building a truck wash. They saw a

82

symmetrical arrangement of columns, wide enough for a truck to drive through, on a commercial corner. So for the first six months everyone in town was guessing what it was going to be. When the arches went up, they really began to wonder."

Richard didn't have the funds to finish the slab and put in landscape irrigation that year, but in the fall dormant season he moved some of the grapevines that had been planted in the original shopping center landscaping in the early 1970s. He selectively thinned the vines along Shaw Avenue, moved the vines to the arbor and planted them next to the columns, and hand watered them until he was able to put in an irrigation system. He wanted color in addition to shade, so he planted climbing roses around the arbor frame.

The following summer the slab was laid, and on July 19, 1984, the market moved under the arbor. It would be three years before the grapevines provided sufficient shade, but it was a good beginning. "I talked my mother into putting her hand prints in the concrete along with the date," Richard said. "She wouldn't put her footprint in; she said it would bring bad luck." Askanoosh Erganian lived to see the grapevines grow to cover the arbor and the Vineyard Farmers Market flourish for ten years under their welcome shade; she died in February 1995 at the age of ninety-four, seventy-four years after arriving in Fresno as a young bride.

The colorful fruit and vegetable mosaics that adorn the four columns framing the arbor entrance were designed and assembled by Chicago mosaicist Cynthia Weiss. Richard contracted with her in the late 1980s to develop a series of decorative mosaics representing the Valley's abundant produce. Describing her designs, Weiss said: "I wanted the beauty of California to be reflected in the mosaics."

At the time the market moved from the parking lot to the arbor, it had evolved to a Saturday market, open from 6 a.m. to noon. Not long after the move, grower Annie Paloutzian suggested adding a Wednesday market. Many growers had such an abundance of produce that they wanted a chance to sell it before it spoiled. The Wednesday afternoon market was added, but it took about three years for shoppers to get in the habit of shopping on Wednesdays in addition to, or instead of, Saturdays. Because of the demands of farming, some growers simply don't have the time to bring their produce to market twice a week.

Fortunately for shoppers hungry for fresh fruits and vegetables, a good number of growers can be counted on to show up Wednesdays and Saturdays year-round. "It's always been a chicken-egg situation: Which comes first, the growers or the customers?" Richard explained. "The growers are waiting for the customers to come first, but the ideal thing is for the growers to make a commitment for an entire season so that customers will know there will be a consistent number of growers there each market day. That gives them more reason to come to the market."

The arbor structure provides parking stalls for twenty-four growers, but the shopper is likely to find more or less than that number on a given day. Late spring and summer are the busiest times; the market is packed with growers selling the freshest fruits and vegetables available. The first cherries, peaches, and nectarines of the season arrived at the market the second week of May in 1995, and berries were not far behind. Fall is also a busy time, as growers appear with tempting varieties of apples, persimmons, nuts, and the first of the citrus crop--along with the last of the late table grapes and the first raisins of the season. As fall moves into winter, some growers take their leave, but a good number remain to supply shoppers with nutritious winter vegetables--broccoli, spinach, chard, cabbage, beets, turnips, all varieties of lettuce and other salad greens--as well as sprouts, whole-grain bread, nuts, dried fruits, avocados, apples, citrus fruits, eggs, tamales, and drought-tolerant plants.

What is a Certified Farmers Market?

To qualify as a certified farmers market, under California regulations passed in 1977 and amended in 1979, produce offered for sale at the market must be sold by the actual growers or their family members or employees. Since 1979, state regulations allow certified growers to sell the produce of up to two other certified growers, provided they are also selling their own produce. That means that to participate, growers have to take time from their farming, or have relatives, employees, or other certified growers willing and able to do so, to bring their produce to the market.

Growers at certified farmers markets such as the Vineyard must have valid certificates issued annually, for a fee, by the county agriculture commissioner of the county where their produce is grown. Certificates list every crop, and the acreage devoted to it, that a grower is actually growing and thus authorized to sell at certified farmers markets.

The California Department of Food and Agriculture began certifying farmers markets in 1977; by November of 1978 there were twenty-five certified markets across the state. According to an article in the February 1979 issue of California Agriculture, ten of these were roadside stands next to fields. Fifteen were public markets, of which three--Davis, San Diego, and San Francisco--were open year-round. By July 1995, there were about 260 certified farmers markets operating across the state, with more springing up each year.

As the only certified farmers market in Fresno, and one of only five in the county, the Vineyard differs from other public markets, city corner produce outlets, and rural roadside stands--except those operated by certified growers--in one crucial respect. The produce sold at the Vineyard, and at other certified markets, has been grown by the person selling it, or by that person's relative or employer. That's why you won't find bananas at the Vineyard Farmers Market. Bananas are a recent addition to California agriculture; they are being grown commercially only at the Seaside Banana Garden, an eleven-acre ranch near La Conchita in Ventura County that has six thousand producing banana plants. If you visit one of the Ventura-area farmers markets, you may find grower Doug Richardson selling his several varieties of bananas--but he probably won't get as far as Fresno.

What does it matter whether or not a farmers market is certified? It depends on your perspective. If getting the lowest possible price for produce is the shopper's primary goal, whether or not the market is certified will make no difference. Many non-certified markets, such as the Arnett-Smith Market run by Florence Arnett Smith and her husband since 1945 at N and Merced Streets in downtown Fresno, offer reliably low produce prices. Florence and her family grow a great deal of the produce offered at that market, and many if not most of the other sellers at that and other open or free markets are the actual growers--but they don't have to be.

That means entrepreneurs are free to scour the countryside for produce bargains to pass on to their customers--still making money in the process. They can pick up overripe bananas from produce wholesalers, surplus tomatoes from local growers, apples that have

84

been in cold storage in other states--whatever might sell. There's not a thing wrong with this, but consumers should be aware of what they are buying.

As difficult as it is to believe, many roadside stands sell produce purchased from wholesalers. That means the customer is getting the same things he or she could purchase in the supermarket. A good way to check whether or not the stand operator is selling home-grown produce, at least in California, is to ask to see a certificate.

Shopping at a certified farmers market is a small but important step in living a life of integrity. By purchasing fruits and vegetables grown in the bioregion, shoppers are supporting growers and enabling them to continue farming and providing for their families and workers. Aside from getting the freshest and most nutritious produce available, often just hours from the tree or field, the customer is participating in a rational and equitable distribution system in which the people who grow the food profit from its sale.

Eating with the Seasons in Your Bioregion

Cruising the produce aisles of a large U.S. supermarket gives the shopper few clues as to what the season might be. While it's true that certain types of produce are more abundant in some seasons than others, for the most part it's possible to find nearly any standard produce item--apples, oranges, tomatoes, lettuce, green beans, potatoes, onions, you name it--365 days of the year. Thanks to the wonders of global marketing, high-speed transit, and commercial refrigeration, tomatoes are never out of season in the U.S. supermarket. You may pay a premium price, but if you crave tomatoes or grapes in February, they're available. People who study such things estimate that the average mouthful of food travels more than a thousand miles from field to table--with many detours along the way.

The certified farmers market is dedicated to the principle of eating with the seasons in the particular bioregion in which the market is located. It makes good sense, when you stop to think about it. Unless you live in the Southern Hemisphere, you won't find grapes at a certified farmers market in February. At the Vineyard Farmers Market, you'll be able to buy one or more varieties of grapes beginning in July and lasting well into December most years. In February, you can satisfy your hunger for fruit with oranges, tangerines, grapefruit, pummelo, and apples, as well as a tempting variety of raisins and other dried fruits. Fortunately for Vineyard Farmers Market shoppers, locally grown greenhouse tomatoes are now available nearly year-round at the market.

Even though he grew up surrounded by produce, Richard says he had never paid close attention to which crops grew where in a particular season. "Certain things were an awakening for me. For instance, in the grocery store we stocked broccoli all year round. I started asking my growers here why there wasn't any broccoli in the market in July. I didn't have my own garden at that time, so I didn't know that broccoli doesn't make it in Fresno in July. All the years in the store when we got broccoli it was an invisible line between when it was local broccoli and when it was coast broccoli. That was an important lesson to me, and the kind of lesson we wanted to get out in the farmers market: living within the seasons of our geography. In other words: Eat foods where they are grown, as they are grown."

Richard generally doesn't encourage growers coming from outside the Central Valley bioregion to sell at the market for two basic reasons: (1) It defeats one of the purposes of having a farmers market: to serve growers from the surrounding community, not 150 miles away, and (2) If a grower is driving two and a-half hours to get to the Vineyard Farmers Market and happens to have a few bad days in a row, he will probably drop out of market. That creates a void and confuses customers who were looking forward to the produce that grower was bringing to the market.

Making a commitment to eat with the seasons where one lives can mean a few menu changes. Instead of tomatoes in that winter salad, we can substitute slices of orange or kiwi--both high in vitamins and locally harvested in the winter months. You won't find iceberg lettuce at the Vineyard Farmers Market in any season, which is actually a very good thing because eating iceberg lettuce is only slightly more nutritious than drinking a glass of water. In place of pallid iceberg, growers offer a palette of brilliant greens loaded with distinctive flavors and vitamins--making salad something more than a vehicle for dressings.

The winter market features abundant Oriental greens for stir frying in creative combinations, as well as hearty root vegetables to be savored greens and all. You won't find as many of these at the market during the summer, but that's when it's so hot you don't want to eat much cooked food anyway--with the possible exception of the succulent sweet corn that tastes like an entirely different vegetable from what one buys in the supermarket. A meal of sweet corn and vine-ripened tomatoes and cucumbers, topped off with melon or peaches, can just about make summer worth waiting for.

Nurturing a Year-Round Market

A visit to the Vineyard Farmers Market during the height of the summer vegetable, stone fruit, melon, and grape season provides ample evidence of the abundant harvest that makes the San Joaquin Valley the nation's premier agricultural region. Valley growers are eager to sell their fruits and vegetables wherever they can. "I get several calls a week all summer from growers wanting to unload excess produce," Richard says. "I'm not at all interested in short-term growers. An important part of our market is the personal relationship between the customer and the grower, and I don't want to start compromising that continuity by adding short-term growers who aren't interested in customer service." He generally advises such people to take their produce to one of the Sunday swap meets, where they'll probably sell more with less competition.

Finding and nurturing growers who can make a year-round commitment, are farming organically, or can bring something different or unusual to the market, is a continuing challenge. "To me, charging rent is secondary to getting a good grower established," Richard explained. "When I find growers that meet these criteria, I'll invite them to try the market free for a couple of weeks to decide if it will work for them.

"I've been working with the growers to develop a sustainable market with produce all year round. It's been a process of trial and error. The grower's attitude and personality, liking to relate to customers, is very important. It's hard to get growers who are 100 percent organic, but they're mostly in transition. My definition of transitional is that they're trying to eliminate chemicals as much as possible. From my observation, the winter market is essentially an organic market because there aren't many pests in the winter.

"My heart is with the organic growers. I'll bend over backwards to help them. Fresno, up

86

to now, is not an organic town." Even though the Vineyard Farmers Market is not fully organic, it certainly beats taking potluck at the supermarket. Shoppers can ask the growers what chemicals they used and when, and what chances there are of having any residue on the produce.

Growers range from home gardeners to farmers with hundreds of acres. Home gardeners are more likely to use organic practices, but giving them space at the market is problematic. They can grow only a limited number of varieties, in relatively small quantities--and mainly in June and July when the market is crowded anyway. "If I let the home gardeners bring in their tomatoes, it puts me in the position of compromising the growers who support the market all year long, even in the winter when sales aren't always that good," Richard explains. He selectively adds home gardeners who are able to provide adequate supplies of something new or unusual not being sold by long-time regular growers, but certified producer certificates have gotten expensive--going from $10 a year when the market opened to $75 a year currently in Fresno County. This extra expense makes it prohibitive for most home gardeners to consider selling at certified farmers markets.

Article on Direct Marketing Conference

provided by OPENAIR-MARKET NET

I began researching farmers' markets in August 1995 in preparation for dissertation fieldwork on marketplace regulation in Oaxaca, Mexico. From September through November, I engaged in participant-observation fieldwork at a weekly farmers' market in suburban Boston. For comparison, I visited other farmers' markets in the Greater Boston area.

Like all marketplaces, farmers' markets in the United States are regulated at many levels. There are rules governing sanitation, weights and measures, where and when vending may take place and what may be vended. A crucial aspect of marketplace regulation concerns the criteria by which vendors are recruited into (or excluded from) the marketplace. This issue has particular saliency with respect to farmers' markets in the United States.

Since the late 1970s, farmers' markets have been supported by federal programs, such as the distribution of food stamps redeemable only at farmers' markets. For example, in Massachusetts, to be classified as a `farmers' market,' and thus receive the benefits of these programs, vendors are supposed to comply with a number of regulations. The Massachusetts Department of Agriculture consults with the Federation of Massachusetts Farmers' Markets, a voluntary, private association, to establish complex guidelines for participation in farmers' markets. Such rules and regulations, however, are not consistent across marketplaces, nor is the enforcement of these rules consistent.

To probe these regulatory issues, I accompanied a local farmers' market manager to the North American Farmers' Direct Marketing Conference in Saratoga, New York from February 22 to February 24, 1996. This years' conference was run by the Executive Director of the Federation of Massachusetts Farmers' Markets. The 1,400 conference participants included farmers, managers of markets, policy makers and government administrators. I observed an all-day seminar for farmers' market managers and a series of presentations and round table discussions focussing on farmers' markets. In each of these sessions, there was active audience participation, during which a number of problems concerning the regulation of farmers' markets emerged.

Among participants of the North American Farmers' Direct Marketing Conference, there was disagreement over the primary function of farmers' markets, and a corresponding disagreement over the function of regulation -- particularly regulation governing the recruitment of vendors into the marketplace. This was not a simple disagreement between farmers and regulators, as the farmers themselves did not agree.

Farmers' markets, by definition, are producers' markets, meaning vendors are also farmers. A major function of farmers' markets is, therefore, to support local farmers. Another major function of farmers' markets is to provision customers, often in urban areas, with local produce. During the conference, disagreement over the relative weighting of these two functions emerged.

Some participants viewed regulation as a means of supporting farmers by ensuring that vendors operate on a "fair playing field." For these participants, regulation was seen as preventing, or limiting, the selling of produce not grown by the vendor. Since resellers (labeled "peddlers" or "hawkers") may be able to provide produce not grown locally, or may be able to undercut the price of locally grown produce, their presence was seen as putting farmers at a competitive disadvantage.

Other participants argued that the primary function of farmers' markets is to serve the customer. For these participants, regulation that discourages resale was seen as "protectionist," creating an "artificial market." Some vendors favored deregulation, which they believed would encourage "free enterprise," allowing vendors to respond with

greater flexibility to the needs of customers. Deregulation, it was argued, would enable farmers' markets to more effectively compete with other produce markets, including supermarkets.

The complexity of these opposing positions has been revealed in legal disputes between vendors and regulators. At the conference, I met regional coordinators of marketplace systems throughout the country. These individuals were willing to discuss various legal disputes between vendors and regulators. In sum, my observations at the North American Farmers' Direct Marketing Conference revealed disagreement among participants over the function of farmers' markets and, correspondingly, over the function of regulation.

While there are some problems concerning regulation that may be unique to farmers' markets in the United States, the conference revealed more general issues concerning regulation of marketplaces. It became clear that regulation may depend on the perceived function of marketplaces. Because marketplaces may serve multiple, sometimes conflicting functions, the regulations promoting these functions may also be in conflict. I plan to pay significant attention to this problem during my dissertation fieldwork in Oaxaca, Mexico.

Excerpt from article by Sharon Young *(used with permission)*

Asked if there was ever a Saturday when he woke up and wished he didn't have to go to market, Angel replies: "I don't think that has ever happened. In fact, I'm always up way ahead of time. Most of the time I wake up at 2:30 on Saturday mornings--automatically." Angel does admit that mornings when he awakens to the Valley's heavy Tule fog he has second thoughts, but they always come to market anyway. One customer who's glad he does is Mary Anne Anderson, who says she's one of the earliest shoppers every Saturday morning. "I could buy carrots at the supermarket," says Mary Anne, "but they wouldn't have been weighed by Angel's pretty wife!" Al Avedekian, who has been shopping at the market since it opened, comments: "I've been there when it was so foggy that only one or two vendors were there." One of those was undoubtedly Angel Garcia.

Connie Goodner, a customer who expressed special appreciation for Angel, commented: "I feel you make friends at the market after years of doing business with them." Long-time customer Joan Westly agrees: "Most of all I enjoy the friendliness of the vendors." One of the friendliest vendors, Angel says: "We have so many friends up here that we know by name. I wish I could know the names of all the people who have been coming for years," he says. "I recognize the faces, but I don't know their names. Everyone wants to be recognized. I think if I knew everybody's name that comes to our stand, it would make them feel a lot better, feel important."

Farmers Working Together for a Fair and Honest Market Place

by Callie Bowdish (12/10/95)
provided by OPENAIR-MARKET NET

True cooperative marketing, where the farmers come together to sell what they grow to the public, could prove to be an important economic strategy for farmers. In California the direct marketing program is still in the birthing stages and now is at a crucial time to determine if it can work. Typically cooperative ventures get taken over by a few strong special

interest people who have borrowing power. The money is used to leverage against the others and the few get a marketing advantage. Skill in growing does not give these special interest their marketing advantage. It is their ability to buy other farmers goods and sell them at a profit. This cycle, of investors out maneuvering farmers in the market place, I feel, can be broken if farmers learn to work together and sell their own produce as a marketing group. I don't see the strategy of farmers coming together as a group as a get rich scheme, but rather as a way for farmers to survive as small scale family and cooperative farms.

Insisting that the people selling in the markets be people who are producing their own crops, I feel, is a key to creating a healthy fair market place. The big question is whether farmers can work together. It is easier for a few moneyed people to use laws with their "gray" areas to their advantage than it is for a large group of people to use laws. It is hard for a group of farmers to develop legislature that help make a fair and honest market place. A relatively small amount of businessmen and people who do not spend hours working in the fields have an advantage over those who work with their hands. It takes a large amount of organization and commitment for farmers to work together and pool their resources.

Farmers work long hard hours. Another difficulty is that it is easier to make a living selling produce that others sold. There is always the temptation for farmers to turn into investors and betray the other farmers. When people sell others products within this cooperative structure they have an advantage of not having to compete against other investors but only other farmers.

In California a Task Force has been meeting to try and determine how to insure the integrity of the Certified Farmers Market program. I feel that the only way this program can come to birth is if farmers learn to understand the necessity of working together. Investors and borrowers can always maneuver the market place to a place where the producers are at a disadvantage. I feel the only way to avoid the farmers from working for and at the mercy of the investors is for the farmers to have their own direct marketing program. Legislature with its laws will mean nothing if farmers don't realize it's importance, and insist that legislature and laws are followed. Laws on paper are nothing if people don't believe they are necessary. Laws on paper are nothing, if people continue to cheat and "sell out" for financial gain at the expense of others. The wholesome, honest farming lifestyle can only be possible if farmers learn how to participate in the market place successfully. My belief is that true direct marketing could help farmers to survive in the market place. The question is whether farmers can learn to work together to survive.

Farmer's Doubters

by Richard Sine
Providedby OPENAIR-MARKET NET

On a sunny day at the little Milpitas Certified Farmers' Market, a guitarist plays cheerful folk tunes. Farmers have driven three hours or more from rural parts of the state to hawk exotic strains of organic nectarines, peas and bittermelon. Local families brave the heat radiating from the concrete at the edge of a parking lot in search of produce that is fresher and riper than what they can find at the local supermarket. But if this is a certified farmers' market -- where growers selling their own harvest are given a reprieve from certain state laws -- then where are the certificates?

More than half of them have not been posted conspicuously at the stalls as required by law. The certificates are supposed to list which products the farmer grows, and can legally sell at the market. Without the certificates, there's little guarantee that a vendor here isn't simply buying someone else's produce at a wholesale market--or even a Safeway--and reselling it as phony homegrown at a higher price. In fact, even with the certificates, there's no guarantee.

The market manager at this particular market says she checks the certificates of the 32 vendors here every day, and most of them are located just out of sight. But a peach grower who goes to six markets a week in the Bay Area says she knows the score. "At all the markets I go to there's always two or three hanging around who don't grow what they sell," she says. Asked to point out who she suspects of peddling at the market today, she blushes like her peaches and declines to make a guess.

There are some 300 farmers' markets in the state this year, more than triple the number of a decade ago. Four new markets are opening this year in Santa Clara County alone, bringing the county's total to sixteen. Farmers' markets owe at least part of their success to their wholesome image. One imagines Farmer John shoving his thumbs into the straps of his overalls, chewing a blade of grass and rocking on his heels as a customer asks him when tomatoes will hit the market this year.

It's true that farmers' markets have been vital to the health of the state's small farms. But the explosive growth of markets across the state has revealed the farmer's trade to be as competitive as any other. Many markets have long waiting lists for farmers hoping for stalls. Excluded farmers accuse market managers of favoritism or even file suit against the markets. ("I've been accused of having affairs with farmers I've let into the market," says one market manager.) Farmers' market associations have been wracked by controversies over how governing boards should be elected. And growing market associations have accused each other of elbowing into each other's territory.

Within the markets, small farmers worry about

how they can compete with larger farms that have caught on to the farmers' market craze. And most of all, they worry about peddlers--vendors who buy produce gleaned from other farmers, packaging houses or wholesale markets and then show up, as faux farmers. Unlike farmers, peddlers can buy and sell whatever crop is cheapest without worrying about being wiped out by cold frosts or pests. They also aren't limited to whatever crops are seasonal. Farmers hope that bills now pending in the California Legislature will help to eliminate the bad apples in their midst. Few of their customers are even aware there's a problem.

The first so-called direct marketing laws, passed in the late 1970s, aimed to limit the influence of the packers, distributors and grocers that divert profits from farmers. (Agricultural authorities estimate that farmers get as little as 25 cents for every dollar spent for their produce at grocery stores.) Certified farmers make a bargain with the state: they avoid regulations on packing and fruit size--expenses incurred by farmers who sell to wholesale markets--in return for a guarantee that they are directly marketing only what they grow. Regulations on quality remain in effect.

But certification laws appear to be loosely enforced. Ibarra-Cruz Organic Farms, a 150-acre farm in Gilroy, is a case in point. For years, Ibarra-Cruz had been selling lettuce without any problems at the Berkeley Farmers' Market, a market with a widespread reputation for eagle-eyed management. Then the floods of 1995 devastated lettuce crops. Lettuce disappeared from farmers' markets, and imported lettuce at supermarkets sold for as much as two dollars a head.

"Ibarra was the only farm at our market to continue to bring lettuce," says Clem Clay, co-manager of the market. "That seemed suspicious, because when we visited them in February there didn't seem to be much more lettuce." At this point, Clay could have looked the other way as customers flocked to his market seeking lettuce. Instead, he sent a letter to farm owner Moses Ibarra demanding an explanation.

When Ibarra did not respond, he asked the Santa Clara County Agricultural Commissioner to step in. In late April, the commissioner's inspection confirmed Clay's suspicions: Ibarra wasn't growing any head lettuce at his farm.

Clay sent out members of his governing committee to other farmers' markets to check up on Ibarra. He says that Ibarra was selling "boxes and boxes" of lettuce at other markets. That was the last straw. In June, Ibarra became the first farmer Clay had kicked out of his market in his three-year tenure at Berkeley.

Ibarra denies that he was reselling produce. He says that at the time Clay had made his accusations, he had started a partnership with a lettuce farmer in Hollister. While the Hollister farmer owned and cultivated the land, Ibarra harvested, hauled and marketed the produce. But Ibarra admits that when Clay first inquired about the lettuce, the partnership agreement was only verbal. (Partnerships are a grey area in farmers' market law. "The line between legal and illegal lease and partnership agreements is very hard to draw," says Mike Thompson, editor of the Farmers' Market Monthly in Inglewood.)

Ibarra blames other farmers at the Berkeley Market for his troubles. He says they were jealous because he was doing brisk sales as a newcomer. "The markets are different than when my father started going in the 1950s," he says. "Back then it was competitive, but farmers always helped each other out when things went wrong. Now, it's a cutthroat situation. When you're at a certain level, there's always people trying to knock you down."

Ibarra says other market managers were satisfied enough with his explanation to keep him in their markets even after they heard of his eviction from Berkeley. The government also let him off. State law allows agricultural commissioners to punish resellers through measures as drastic as revoking their certification. But neither Santa Clara County, where Ibarra's farm is located, or Alameda County, where he sold the goods, took any action against him. Ed Williams of the state Department of Agriculture estimates that commissioners suspend or revoke only eight to 10 certificates a year.

Francis Schmidt Leon, who commutes to her family farm in Selma from her home in San Jose, says some market managers ignore peddling so they can carry a wider variety of produce."They might need that product at the market to get more customers there, or they just need to fill the stall space," she says. The temptation to allow peddling can be strongest during the winter months. In winter, supermarkets can carry produce from around the world but markets are limited to what farmers can grow locally. But farmers' market ideologues say it is ecologically sound to get consumers in tune with the seasons and the local market supply. They say it supports local farming and farmland preservation and saves the petroleum and materials used in packing and shipping to distant markets.

Even the most honest market manager may find it difficult to seek out peddlers because of loopholes in the certification process. Current laws make it possible for farmers to "carry" other farmers' certificates and sell their produce. The laws were designed to allow neighboring farmers to assist each other, but they have opened the door to peddling. "You'll ask a farmer where some produce on their table came from, and they'll pull out 15 or 20 certificates," says Williams of the agriculture department.

Typically, market managers who sense shady dealings will request an investigation by the county agricultural commissioner. But certification programs are underfunded when compared to other programs that occupy the commissioners' time, like pesticide control and inspection of imported fruit. Santa Clara County now charges farmers a yearly $15 certification fee, regardless of farm size. As a result, county biologists can't personally inspect all farms every year to ensure that a farmer is growing what she claims.

Even a visit to a farm can't insure against fraud. Lucero says a county biologist visited him in February this year, before he had even planted his summer crops. So he had to guess how much he would grow in summer, and the biologist approved his guess on the certificate. "It's a brief and not a very critical check," Lucero says. "Unfortunately, it leaves it kind of open for anyone who wants to cheat."

Farmers' Market Produce Prices: A Multivariate Analysis

by Bill Blake (UC Cooperative Extension)

November, 1994

Abstract

Consumers shop at farmers' markets for many reasons, one of which is that a market can be an "event." A market can provide a focal point for a community or foster interaction between people. From an economic perspective, consumers who attend markets for these benefits should be willing to pay more for the products at the market, in effect purchasing their participation in the event. Therefore, markets intended as community events and those intended as sources of low-cost food may have different design and marketing requirements. The first step to examining this question, however, is to determine how much difference the "event" status really makes. This study indicates that farmers' markets across the board offer lower-priced produce than neighboring grocers, but that the event status of markets seems to affect prices. However, this study also finds that econometricly quantifying influences on market prices is problematic.

I. Introduction

A farmers' market is both an event and a point of sale. As an event, it provides a focal

92

point for a community, an opportunity for direct exchange between growers and consumers, and a place for socializing. As a point of sale, according to conventional wisdom, it offers better quality produce at lower prices because farmers can market directly to consumers. Produce from farmers' markets has been shown to be preferable to produce from grocery stores (Sommer, et al., 1982), and this improved quality should tend to increase prices. The additional utility of participating in the "event" should also push prices higher. On the other hand, the market is less convenient than a grocery store, and the produce is not as clean, uniform, or cosmetically controlled. These factors exert a downward pressure on prices.

This paper tests the hypothesis that prices at farmers' markets are lower. Since there are forces working both to raise and lower prices, this is an empirical question. The question is interesting because once it has been determined whether prices are lower or higher, one can address the importance of the utility that consumers derive from the "event" of the market.

In previous work, Sommer, et al. (1980) looked at markets in Northern California. In a survey of nearly 358 items over the course of several months, they determined that prices were lower by 37% for vegetables and 39% for fruits. Their article also rehashes the result of divers other studies, mostly from the East Coast, which indicate savings ranging from 8% to 50%. They point out, however, that these other studies are not comparable, and that they suffer from ambiguity around the term "farmers' market." The present paper has the convenience of using the same definition as the Sommer, et al. survey, that of a market certified by the California State Department of Food and Agriculture.

While the Sommer, et al. survey was extensive and complete, it did not attempt a multivariate analysis. A great number of factors influence grocery store prices, and most likely also affect prices at markets. These include size of the store, amount of local competition, population, local median income, etc. (Cf. for example Maley, 1976). This study employs a multivariate regression analysis so that the influence of each of several variables could be tested and quantified.

However, the effect of marketing directly could not be separated from other aspects of the farmers' market. The market is simply a different product than the grocery store. Freshness, taste, and the experience of the event are all interior to the produce at a market. Convenience, sterility, cosmetic quality, and one-stop shopping are likewise interior to the grocery store's produce. This paper does not simply compare produce at one final purchase point with produce at another. It compares farmers'-market-produce to grocery-store-produce, with the many sociological and psychological attributes that both composite goods entail.

II. Data and methods

The analysis is conducted on a set of cross-sectional data collected by the author over an eight-day period in mid- February 1994 in five Northern California cities which have regular, year-long, certified farmers' markets. Those places were Davis, Sacramento, San Francisco, Berkeley and Marin. Prices were collected for seven widely available vegetables or fruits: greenleaf lettuce, red potatoes, yellow onions, bunched kale, Fugi apples, oranges, and kiwi fruits. Prices in each locale came from the farmers' market and from three neighboring stores, and were differentiated into certified organically grown and otherwise produced. An item could have a theoretical total of 26 observations (8 from markets and 18 from stores, or 19 conventional and 7 organic). Actual totals ranged from 18 to 25 observations.

Individual prices for each item were then divided by the sample mean for that item's prices in a single locale to create an indexed price:

$$z_{ij} = P_{ij}/P_i$$

where z_{ij} = indexed price of ith produce at jth point of sale

P_{ij} = ith price at jth point of sale

P_i = sample mean of an item's prices in a particular city

Data were collected for a number of independent variables. The variety (var) of produce was measured by counting the number of different produce items offered. As is often the case in dealing with produce, this led to problems of definition. In this survey as in Sommer, et al., 1980, if a seller made a distinction it was honored in the variety count. Thus, bunched beets and loose beets are two items, large oranges and extra large oranges are separate, and some stores had nearly a dozen apple varieties. Number (num) was meant to capture the effects of competition by measuring the number of competitors. For stores, managers were simply asked how many stores they competed with. For markets, this was defined as the number of stalls offering five or more different items. The number of shoppers (shop) per day controls for customer traffic. A higher volume of traffic can account for price differentials, both between stores and between the markets (some of whom have very high volumes, especially in the peak summer season) and grocery stores. The size

of the selling area is measured in square footage, and is included to test the relationship between the physical size of the store or market and prices.

There is a dummy variable for organically grown (org) produce, 1 if yes, 0 otherwise. A second dummy variable records the effect of farmers' markets (mark) on prices. Six dummy variables, v1 to v6 capture any variability due specifically to any of the seven vegetables. A final variable questions whether a store being a co-operative (coop) has any effect on prices. Guy and O'Brien (1983) found a significant difference in prices between multiple retailers and co-operatives on the one hand, and affiliated and independent retailers on the other. In fact, a difference in prices for farmers' markets could be seen as simply a special case of the effect of corporate structure on prices.

The model to be estimated is given by:

(1) z_{ij} = z_{ij}(var, num, shop, size, org, mark, coop, v1-v6)

where:

var = number of different produce items available at store or market

num = number of competitors

shop = number of shoppers per day

size = square footage of market or store

org = dummy variable: 1 if organically grown, 0 otherwise

mark = dummy variable: 1 if a farmers' market, 0 otherwise

v1 = dummy variable: 1 if lettuce, 0 otherwise

v2 = dummy variable: 1 if potatoes, 0 otherwise

v3 = dummy variable: 1 if onions, 0 otherwise

v4 = dummy variable: 1 if kale, 0 otherwise

v5 = dummy variable: 1 if oranges, 0 otherwise

v6 = dummy variable: 1 if apples, 0 otherwise

coop = dummy variable: 1 if co-operative store, 0 otherwise

The expected signs on the coefficients of the variables are as follows:

var(+), num(-), shop(-), size(-), org(+), mark(-), v1(0), v2(0), v3(0), v4(0), v5(0), v6(0), coop(0)

The variety of items should increase the overall prices for several reasons. First, variety of offerings can take the place of low prices as an inducement to shop at the store or market. Second, the extra variety will tend to come from specialty items which are purchased less frequently and which may also (but not necessarily) be more fragile. These characteristics will lead to higher produce losses, which must be compensated for by higher prices. The number of competitors should lower prices as they increase the supply and compete for business. The number of shoppers could raise prices by increasing demand, but will more likely allow for greater sales per sales area and per hour worked, reducing the amount of fixed cost (such as store rent) that must be covered by each individual sale. Size should have a negative effect on prices because large size allows for bulk purchases and greater turnover for stores, or increased mechanization of farm operations

and efficient use of management abilities for farmers. Organic produce is generally more expensive than conventional produce, and results from this study are not expected to be different. The variable for cooperatives is not expected to have an effect on prices, for reasons discussed below. Finally, the type of vegetable (v1 - v6) should not affect the overall result.

Early models with this data were heteroskedastic. This occurred either if the data was indexed with a sample mean taken across all locales, or if a single regression was attempted for all the data. This heteroskedasticity was determined by a chi-squared test. Despite the lack of a uniform variance, a Chow test was performed on the data; not surprisingly, it showed the data points breaking down by locale. Part of the problem was that price differentials between the farmers' markets and the stores did not correlate with variables collected to describe the different locales, i.e. population sizes and income levels. Tastes and preferences marginally related to these simple statistics or variables unmeasured here greatly affect pricing strategies. I chose therefore to perform a separate regression for each locale, with prices indexed as indicated above.

Furthermore, several variables were eventually omitted from the final models. The dummy variables for the individual vegetables (v1 - v6) had to be dropped in order for the several estimations with the reduced number of observations to solve. The variable num was found to co-vary significantly with everything but income and store size. It was therefore dropped because it over-specified the model. This action had almost no affect on the goodness of fit, but did improve the t-ratios of other variables. An additional reason for it being dropped was that its specification was problematic. For a store, the variable was defined as number of competing stores; for a market, it was number of stalls selling more than just a few items. The intention was to capture the effects of competition, for it has been shown that concentration of the retail grocery industry in an area leads to higher prices (Marion, et al., 1979). However, competition between stalls and that between stores is not directly comparable. A variable to capture competition would need to be more complex.

Two other omitted variables were shop and coop. Shop had significant covariances with nearly everything, so it was dropped as a redundant variable. The unfortunate consequence of removing shop is that the utility derived from attending the farmers' market because of its status as a popular community event is not separated from the utility derived from purchasing fresh produce direct from the grower. The variable coop was abandoned early. First, there were only two co-operatives in the data set, out of fifteen total stores, so the number of observations was small. Second, casual observation by the author and others has failed to note much difference between co-ops' prices and those of grocery stores. Since no significance was anticipated, when early estimations ascribed to coop a low t-score as well as a low coefficient, it was removed from further regressions. The removal had almost no effect on the estimation.

III. Results

The results of the regressions for the individual locales demonstrate that the structure of the model obviously changes by area. This confirms the problems of correlating the price differentials to the statistics used to describe the locales, i.e. population and income level. Variables which are highly significant in one locale cease being significant in another. Every variable but org has both positive and negative signs, and that variable varies in magnitude. Even the most important variable for the purposes of this paper, mark, changes sign, size, and significance. These results demonstrate the complexity of pricing strategies, in particular those of farmers' markets.

That said, it is important to note that farmers' market prices are generally lower. For four of the five locales, the coefficient for the variable mark is negative and the absolute value of the t-ratio is greater than unity; for two of them, the t-ratio is significant at the 5% level. Only one does not show this tendency, and that one indicates parity between market and store prices. The results, in fact, confirm the impressions that this author has formed from shopping and working at the different markets.

At Marin, the market is held in the parking lot of the Civic Center designed by Frank Lloyd Wright. It dedicates a lot of space to non-farmer vendors selling artsy hats, specialty vinegars and dressings, and other craft items. Many of the shoppers are economically comfortable. They enjoy the bazaar atmosphere or European flavor of a market, and running into each other there seems to have a certain cachet. Consequently, farmers are not as hard-pressed to offer bargain prices to assure a clientele, because shoppers derive utility from aspects of the market other than the low prices. The contrary case is Sacramento, where the Sunday market takes place in a parking lot under the freeway. Nearly the entire space is dedicated to farmers, the exceptions being the fish vendors and the two or three local bakeries which have stands. Shoppers tend to be less well-off economically, and many are from the ethnic communities of the city. They are concerned with freshness and price, as well as with finding produce items that grocery stores might not carry. It is therefore not surprising that the produce was much cheaper at the market, and that the t-ratios showed very high significance. In this vein, it is interesting to note that the Sunday Sacramento market does not have a single certified organic farmer.

IV. Conclusion

Perhaps an economic model of farmers' market prices needs to include more variables to be generally applicable. What is more likely is that the effect of the market on prices depends on the character of the locale, on the weight it places on direct marketing, the event of the market, and price savings. As for the central question of the paper, whether prices at farmers' markets are lower, it seems that they by and large are lower, though by how much will vary according to area. An interesting direction for more study would be to amass new time-series data to compare to the set from Sommer, et al. (1980), to try to uncover changes which have occurred in the interim. Furthermore, attempts should be made to identify sociological variables affecting farmers' market prices, especially shoppers' perceptions and their reasons for participation. These could help clarify the complex good which people are "purchasing" when they shop at a farmers' market. This information can help community and city planners articulate their goals when starting a market (is it to be a source of low-cost food? a community center? both?) and deciding how to design it, where to hold it, and what vendors to allow participate.

Author Biography

William H. (Bill) Blake III first noticed market pricing differences when working on an organic farm. A Virginia native, he earned his M.S. at the University of California, Davis in International Agricultural Development. He is currently involved in several projects for UC Cooperative Extension (UCCE) and UC Davis, and is the main author of Making the Connection, a UCCE handbook for Community Supported Agriculture.

Appendix B

Selected News Articles and Press Releases

1.

STAUNTON/AUGUSTA FARMERS' MARKET GROWTH EXCITING

VENDORS, CUSTOMERS COMMENT ON DELIGHTFUL SATURDAY MORNINGS IN DOWNTOWN STAUNTON

Staunton, VA - It was barely past sunrise on a recent Saturday Morning when Robert Brown eased into the Wharf Parking Lot in downtown Staunton. The produce he was bringing to market was absolutely fresh. In this, his second season as a vendor at the Staunton/Augusta Farmers' Market, "Farmer Brown" had arrived on this particular morning with sweet corn, vine-ripened tomatoes, green beans, rhubarb, zucchini, patty pan squash, home-made jellies and jams, and fresh muffins from his home near Mt Sidney.

The Staunton Augusta Farmers' market is now halfway through it's third season of exciting growth. From an eager handful of original farmers, the market has grown to a roster of over 30 local vendors.

Regular customer Peter Olsen says "Everything is so fresh and keeps so much longer than what I can get at the store. Also, it's become almost a social occasion, where I see all my friends at the market on Saturday mornings. Lot's of times, I come even when I don't need anything." Clearly, Saturday mornings at the market are an enjoyable experience.

According to Market Master Marilyn Young, "At this point in the season, we are way ahead of last year. We are really excited about the financial outlook of the Market." This is the peak of the fresh produce season, and every Saturday morning is a "beehive of activity" around the vendor's booths.

There always seems to be a commotion in front of Sam and Sarah Ann Yoder's market stall. All the bustle is caused by customers ogling the fresh apple dumplings, sticky buns, German chocolate cake, rhubarb pie, shoo-fly pie, dietetic applesauce cookies, zucchini bread, wholesome harvest bread, white bread, salt-rising bread, braided bread, and peanut butter cookies. Most of the Yoder's customers are regulars, and if you taste any of Sarah Ann's baked goods, you'll be a regular too!

The Staunton/Augusta Farmers' Market is a "Producer's Only" market. Producer is defined as the "person that grows or makes the product and may include the producer's immediate family members, partners, or employees or local cooperatives upon review. Selling of items purchased from, or provided by, another market is not permitted." The reason behind this market rule is that customer's have repeatedly asked for "local produce grown by local farmers". If you want grapefruit, you are going to have to try someplace else than the Wharf Parking Lot.

However, if it's organically grown salad greens, you need to see Mark Van Lear's booth. Almost every Saturday, you will see fresh and chemical-free lettuce, Italian parsley, sprouts, endive, chard, and cut flowers. And if you need something to spruce up the dining room table while enjoying Mark's salad crops, pick up a fresh bouquet of colorful cut flowers.

Market Master Marilyn Young is also excited about the increase in the number of vendors from the region coming to downtown Staunton on Saturday mornings to sell their fresh farm goods. "We now have folks coming in from Fishersville, Swoope, Burnsville, Waynesboro, West Augusta, Afton, Mt Sidney, Middlebrook, Ft. Defiance, Stuart's Draft, Bolar and Lexington." It is obvious to customers that there is a wealth of fresh fruit and vegetables grown in the Shenandoah Valley, and on Saturday mornings, it all seems to head for downtown Staunton.

Steady patron Joan Rice of Mt Sidney told Augusta Country "I've been coming to the market regularly for about a year now. The one reason I come is simple . . . fresh produce. I love to see a variety of food, and you can

surely see it here. I lean towards organic grower's produce" which can be found regularly at the market. She continued by saying "One thing I'd like to see is crafts and art work by local artists also. It might help attract more customers to downtown. But it would have to involve local people only".

Each vendor at the market is required to complete a "Producer's Certificate" before selling at the market. New vendors can request forms from the Market Master Marilyn Young on location at the Wharf Parking Lot, or by stopping by the Staunton City Treasurer's Office in Staunton City Hall, the Augusta County Farm Bureau, or the Augusta County Extension Service Office in Verona.

Merritt and Linda Liptrap have become anchors of the shaded sidewalk on Saturday mornings. The Liptrap family produces local honey products, including flavored and regular honeys. Linda also sells homemade apple butter, rose geranium jelly, cinnamon basil jelly, Queen Anne's Lace jelly, and a broad variety of fruit and herbal vinegars. The Liptraps were joined on a recent Saturday morning by grand-daughter Heather Hewitt, age 10, who was displaying with pride her carrots, beets, and fresh hand made muffins.

Market rules, drawn up by the Market Committee and vendors themselves, state that "Children under the age of fourteen shall not be a vendor unless accompanied by an adult responsible for the child's conduct and safety".

This rule is well known to one two-year veteran of the market, Ellie Frazier, who just turned 8 years old in June. "Be sure to say that Helen Keller and I have the same birthday" Ellie said exuberantly. Ellie lives in Staunton, and grows Queen Anne's Lace, oregano and mint for the market. "I plant the seeds in the Spring, and mom helps a little. This week I have a homemade herb mask that I want to sell, but nobody seems interested yet." When this reporter asked if perhaps the herbal mask could be used to help in revitalizing facial skin, Ellie looked at me for a minute and proclaimed "Noooo. You wear it just for fun." Oh. You learn something new every day, and this vivacious 8 year old had just reminded me that the Farmers' Market has everything to do with fun.

This year, the market opened earlier than in past years. April 22 was the "opening day", with business as usual continuing through the year until the last Saturday in October. Marilyn Young advises that "traditionally, we have seen the lowest sales figures at the beginning and end of the season; however, our operating expenses remain the same during these periods". This year, market vendors are hoping to see the summer rush carry over right into the Fall. Many have planted specific crops for the Autumn market, such as pumpkins, winter squash, Fall berries, etc.

Matt and Linda Cauley drive in nearly 50 miles every Saturday morning from Millboro Springs in Bath County. In their third year of market farming, the Cauleys can be counted on to have fresh seasonal produce, such as salad greens, spinach, cabbage, pickling and slicing cucumbers, snow peas, green beans, sweet peppers, hot banana peppers, dill and other herbs, endive, eggplant and beets. Matt advises Augusta Country readers "we have planted a full crop of salad greens for this Fall", so if you want fresh vegetables from a family farm, Matt and Linda are at the market every Saturday morning.

There are two rows for vendors at the Wharf Parking Lot every Saturday morning. The first row along the brick sidewalk is reserved by farmers who paid a fee at the beginning of the year. Other spaces are assigned on a first-come, first served basis each Saturday morning. Vendors start arriving shortly after 6:30 and open for business at 7 AM, rain or shine.

Perhaps the best reason to visit the Staunton/Augusta Farmers' Market next Saturday morning is defined by steady customer Jenifer Bradford. "There are always a lot of good little things to eat."

2.

Variety is the Keyword for Local Farmers' Market

Staunton, VA - Among the many vendors on a recent Saturday morning, here is a sampling of items available at the Staunton/Augusta Farmers' Market:

- Fritz Flower brought in from Churchville a variety of indoor plants, including ferns, ivy, Dieffenbachia, Chocolate Soldier Episcia, blood plants, chicken gizzard, bromeliad, and bird's nest ferns.

- Jim Chaffins from Staunton sold an array of fresh local garlic, including varieties such as Elephant, Silverskin, and Roja garlic. Jim is in his third year at the Staunton/Augusta Farmers' Market, and also sells seasonal vegetables.

- Kristen Rogers of Buffalo Creek, near Lexington, had a gorgeous collection of homemade vinegars, including Tarragon, Sage, Rosemary, and Country Italian vinegar. She also had for the customers a table full of appealing fresh baked goods, such as whole wheat rye braided bread, pumpernickel bread, blueberry muffins, and banana-chocolate chip muffins. Kristen is a member of the "Virginia's Finest" program, and sells regularly at the Wharf Parking Lot on Saturday mornings.

- Brothers-in-law Mark Hanger, a volunteer firefighter from Summerdean, and Robbie Cline, a greenhouse grower from Middlebrook, had enough summer produce to satisfy anyone's desire for fresh summer vegetables, consisting of cucumbers, yellow squash, broccoli, green beans, beets, tomatoes, and picturesque cut flowers.

3.

FARMERS' MARKET ACCELERATES TO NEW HEIGHTS

Staunton, VA - The increasing popularity of the Staunton/Augusta Farmers' Market continues to overwhelm vendors who sell there. Market Master Marilyn Young reports "This year we seem to be breaking every record there is. Our farmer participation is up, our weekly sales are up, and excitement about the market continues to expand. In 1993, the first year of the market, our goal was to have one or two days each season when sales exceeded $2000.00. Now our market, as a whole, is averaging almost $3000.00 per market day, and periodically we hit $4000.00! We are not sure where the top is, but we know that market consumers seem happier than ever, and our farmers are all beaming this year."

Farmers' representative Jeff Ishee said "This market season has been a great one for every vendor, and we still have several weeks left. There have been some Saturday mornings at the Wharf parking lot where we were completely besieged with customers. Of course, the weather has been ideal for vegetable and fruit production. Thankfully Hurricane Fran came just after the sweet corn harvest."

Sales figures for the market this year prove the acceptance of genuine farmers' markets like Staunton/Augusta, where each vendor must produce his/her own product, hence the term "producer's only" market. "There have been several attempts locally to get in on the action caused by the popularity of farmers' markets nationwide" said Ishee. "The USDA reports that there has been a 40% increase in the number of local farmers' markets in the last two years alone. In Staunton, our market is so popular that area consumers are spending an average of $800.00 per hour some mornings, and of course, there are no middle men to siphon off the farmers' profit. We have several customers who do a major percentage of their grocery shopping at the market getting fresh bread, produce, meat, eggs and other farm products; therefore, we have started providing shopping baskets as a convenience for the customer."

Ishee encouraged other area farmers to take part by saying "Right now, the demand for fresh local produce far exceeds the supply, and area farmers can cash in on this demand by planning now for next years market season. There is no long term commitment to be a vendor at our market. You can come every Saturday of the 7 month season, or you can come once or twice. It's up to the farmer. We'll provide the market if you can just supply the farm products.

The Staunton/Augusta Farmers' Market is open Saturday mornings at the Wharf parking lot in historic downtown Staunton from 7:00 a.m.until noon, rain or shine. The market season will continue through the last Saturday in October.

4.

Staunton/Augusta Farmers' Market
P.O. Box 58, Staunton, VA 24401

FOR IMMEDIATE RELEASE
Contact: Jeff Ishee
Phone: (540) 886-8477 - Farm Office
             (540) 480-9866 - Car
FAX: Not yet
E-Mail: Hah!

SWOOPE TEEN FARMER TAPS INTO GROWING MARKET

Staunton, VA., May 13 -- "It was just a matter of fiddling around until I got it right" says 14-year-old Daniel Salatin of Swoope. The young farmer was talking about his newest farm enterprise, which is making maple sugar donuts from scratch.

Daniel explained "The first thing I did to make these donuts was to tap into 3 big old maple trees on the farm. Dad and I got about 40 gallons of sap from the trees, boiled it down, and ended up with about a gallon of syrup. Then last week, I made some cake donuts. It took a few tries to get the recipe just right, but the test batches went pretty quick between my little sister Rachel and I. We made a glaze with the maple, dunked the donuts in that, and we were finished. While the donuts were cooling on the table, a regular egg customer dropped by the farm, sniffed the aroma coming from the kitchen, and asked what we were making. Before I knew it, I had made my very first maple donut sale - without even really trying."

The young entrepreneur joined his father at the Staunton/Augusta Farmers' Market Saturday morning and sold his maple donuts along with the family's eggs and USDA inspected beef and pork. With a big smile beaming from his face, Daniel reported "It was great this week at the farmers' market. My donuts sold out before 10 o'clock."

Excelling in farm enterprises is nothing new to the young man. Growing up on his family's diversified poultry/beef/pork/tree farm near Little North Mountain, Daniel wanted to raise rabbits when he was 8 years old. He is now producing more than 1,000 rabbits a year for area clients. He calls his customers, keeps his own records, and does all the daily production work.

"The business is his" says father Joel Salatin. "He is learning that agriculture can be emotionally and financially rewarding. In fact, he is now beginning to speak at agriculture conferences explaining how he has developed probably the only commercial rabbit flock in the country genetically selected for forage metabolism. We are very proud of him."

Daniel's fresh maple donuts are the newest farm product at the Staunton /Augusta Farmers' Market. Market Master Marilyn Young reports that with the addition of several new vendors such as the Salatin family, the market is having a record year so far. "We have had astonishing sales figures the first five weeks of this year. In fact, our volume at the farmers' market has doubled when compared with the same time period last year. The way things are growing, I suspect it won't be long until our farmers are doing three times the volume they had last year. People seem to have suddenly discovered our market. On a typical Saturday morning, during the peak hours of 7 - 10 am, it is usually elbow to elbow."

Farmers are usually inclined to avoid crowds; but when it's a crowd at the Staunton/Augusta Farmers' Market, young Daniel Salatin doesn't mind that at all.

The market is located downtown in the historic district of the Wharf Parking Lot, and is open rain or shine every Saturday through October. Hours are from 7 am - noon.

5.

Staunton/Augusta Farmers' Market
P.O. Box 58, Staunton, VA 24401

FOR IMMEDIATE RELEASE
Contact: Jeff Ishee
Phone: (540) 886-8477 - Farm Office
                (540) 480-9866 - Car
FAX: Not yet
E-Mail: Hah!

FARMERS' MARKET TEEN RECEIVES STATEWIDE AWARD

Stuarts Draft, VA., June 5, -- "It's been a really busy spring" says 18 year old John Lindeman of Stuarts Draft. "My plants have been selling very well at the farmers' market, I got a promotion at my job recently, and graduation (from high school) is only a few days away."

Indeed, this young man has been busy . . .

very busy.

A typical day for John starts early in the morning as he tends to his greenhouse crops before going to school. After a full day of classes, he rushes home to do watering, then goes to work at Tastee-Freez in Stuarts Draft where he serves as Assistant Manager. Getting off shortly before midnight, he then hurries home and does homework and anything else that needs to be done.

"John is a very busy young man" says his mother. "The greenhouse business is his first love. He handles it all by himself - paying the bills, ordering supplies, making lists, and selling the plants to customers here at the Staunton/Augusta Farmers' Market. He is totally in charge of the business."

The Lindeman family moved to Augusta County from Long Island, New York just five years ago. Considered by his 5 brothers and 2 sisters to be "the family gardener", John started growing peppers and tomatoes when he was younger. "Now I grow herbs, geraniums, begonias, salvia, ageratum, and vegetable plants" conveys the young entrepreneur. "Dad built a greenhouse, but it blew away in a storm. Now we have a 16' x 32' house with propane heat and all the right equipment." Addressing how his business has grown, he says "a few people came out to the house last year to buy plants, but now most of my business is done at the farmers' market.

"At the moment, the greenhouse work is kind of light, compared to a few months ago. This allows me to put in 50 - 60 hours a week at my Tastee-Freez job. At the greenhouse, watering is the major job now. It was busy, though, all winter long. Last November is when I started some of the crops for this year.

"Basil is my favorite thing to produce. I grow many different types of basil, including purple ruffles, sweet, cinnamon, licorice, and compact varieties. I also grow French, English, lemon, and woolly thyme."

The senior at Riverheads High School was recently rewarded for his horticultural and marketing efforts by being chosen to receive the Outstanding Youth Award from the Virginia Farmers Direct Marketing Association. He and Market Master Marilyn Young have been invited to Virginia Beach on July 30 where he will be acknowledged by the organization.

Lindeman says "I really like to work with customers. Eventually, I would like to own a restaurant." The combination of this young man's green thumb, marketing skills, and adeptness at hard work almost guarantee him a successful future.

The Staunton/Augusta Farmers' Market is located downtown in the historic district of the Wharf Parking Lot, and is open rain or shine every Saturday through October. Hours are from 7 a.m. - noon.

6.
RADIO *PSA*
(PUBLIC SERVICE ANNOUNCEMENT)
FOR STAUNTON/AUGUSTA FARMERS' MARKET

The City of Staunton and Augusta County take pleasure in announcing that the Staunton/Augusta Farmers' Market will open at 7 AM on Saturday, April 13th, at the historic Wharf parking lot in downtown Staunton.

Dozens of local farmers will be there with fresh farm and kitchen products, including seedlings, house plants, baked goods, honey, eggs, jams & jellies, and locally produced beef and pork.

Farmers from Churchville, Fort Defiance, Stuarts Draft, New Hope, Fishersville, Middlebrook, Millboro Springs, and many more area communities will be at the Wharf parking lot in downtown Staunton bright and early on Saturday, April 13.

That's Opening Day of the 1996 Staunton/Augusta Farmers' Market. 7 AM 'til noon Saturday, April 13th.

Local farmers wish to thank WSVA for this public service announcement.

7.

RUSSIANS LOOK AT AGRICULTURAL ALTERNATIVES IN AUGUSTA COUNTY

DELEGATION FROM BELARUS VISITS SUSTAINABLE FAMILY FARM

by Jeff Ishee

Middlebrook, Virginia - The questions came in rapid-fire succession. "Who maintains the roads on this farm. Who guards the farm at night? How do you cut feed costs so dramatically? Are you allowed to travel? What do you want to make money for?"

Joel Salatin, innovative Augusta County, Virginia farmer, answered the onslaught of questions methodically and precisely. But the typical questions heard on most farm tours, such as "How do you do this?" never transpired. This group wanted to know "Why? Why do you do this?"

This was not a typical farm tour sponsored by the local extension office. This was a direct inquiry from the government of Belarus, who invited themselves recently to the Swoope farm of Joel & Teresa Salatin.

The phone call came about five days in advance. On the line was Valentin Rybokov, representing the Ambassador of the Republic of Belarus to the United States. He asked "Is it possible to tour Polyface Farm with a small delegation from Belarus this Saturday? We have heard from sources here in Washington that methods and ideologies utilized on your farm may have applications in our homeland. Mmmmm, yes. This Saturday."

It was apparent to the Salatins that there was a sense of seriousness in the request. This presumption held true, for in Belarus, farming is in a state of crisis.

BACKGROUND OF FARMING IN BELARUS

Until 1991, when Belarus declared independence from the failed Soviet Union, all agriculture in the country was executed as one collective farm. In fact, two of the gentlemen visiting Polyface Farm recently were former chairmen of the state controlled Collective Farm, both being agricultural scientists with advanced degrees in agronomy. The transition from collective farming - which was used in most communist nations - to privatized farms has been a difficult one for the inexperienced government to accomplish. Currently, there are both collective and privatized farms in the country, with the goal of having all farms become privately held.

In mid-1992, Supreme Soviet chairman Shushkevich proclaimed "we have decided in principle that we are moving toward a market economy, but we are not doing it in the way our western and eastern, or northern and southern, neighbors do. We have decided that we will have a Belarussian way, . . . using our Belarussian intellectual power."

That same year, the value of agricultural commodities traded shrank by 26 percent. The government in Minsk, the capitol city of Belarus, was forced to declare a "state of emergency" in agriculture. Many nations willingly came to the rescue, with assistance coming from several sources. The focus of U.S. aid was the provision of humanitarian and agricultural support for a balanced transition to a free market economy. In FY 1992, the USDA provided $24 million in concessional loans to Belarus to purchase U.S. agricultural products and about $15 million in commodities. In FY 1993, the USDA provided corn valued at $19 million and over $20 million in other commodities. Through December 1993, the U.S. has provided about $140 million in assistance to Belarus (not including nuclear weapon dismantlement programs, which added $75 million via the Nunn-Lugar Act).

U.S. advisors are currently assisting Belarus with training in the fundamentals of agribusiness and a free market system. For decades, the Collective Farm has been one of specialization, with farmers accomplishing one task before turning a product over to another member of the joint agricultural effort. The nation evolved into assembly line agronomy, becoming essentially a nationwide factory farm. The transition to diversified farms where the owner has sole control over his operation is proving difficult and perplexing to the people on the land.

Along with the agricultural "state of emergency", an additional financial burden on Belarus has been imposed by the necessity of dealing with the nuclear disaster that occurred in neighboring Ukraine. Seventy percent of Chernobyl's fallout zone occurred inside what is now Belarus. The combination of economic, agricultural and elemental pressures of this kind have prompted Belarussian officials to explore several different alternatives in its transition to privatized farming and a market economy.

INQUIRY AT POLYFACE

The leader of the visiting delegation was Semyon Sharetsky, Chairman of the Supreme Council of the Republic of Belarus. Mr Sharetsky, elected in December 1995, is also the current leader of the Agrarian Party in Belarus. One question that Chairman Sharetsky asked during the tour at Polyface Farm defined the difficulties that Belarus is having. He bluntly asked "What do you want to make money for? Do you want to expand to a bigger farm? Do you want to buy a fancy house in Washington, D.C.? Do you want to go on a nice vacation?"

It was evident that Joel Salatin was struck by the frankness of the question. Here he was explaining to a former chairman of the Belarussian Collective Farm why someone should farm privately. He was explaining the desire to farm, which apparently is not manifest in the population of Belarus. Salatin answered the question by saying "Why would I want to move to a fancy house in a big city? I'd much rather make refinements to my farm and live happily here. Look around you. We have a healthy and thriving farm here, with a large, and content, customer base. We have a wonderful environment to raise children. What more could a man want?"

Which led to another question from Chairman Sharetsky: "A customer base? What is that?" Salatin responded by explaining that he has people who come directly to his farm and purchase beef, poultry, eggs, pork, and other products generated on the farm. The customers sign up in advance to buy his products, and consequently, most of the farms yield is sold in advance. Another outlet available to him is the *local farmers market*, where - again - he sells directly to the consumer.

This factor was visibly startling to the Russian officials. They asked "Are there no middlemen?" Salatin articulated that, except for the USDA approved processing plant in a neighboring county, "No. There are no middlemen."

Mechislov Guirut, member of the Presidium of the Supreme Council of the Republic of Belarus, then queried further by asking "Do you mean to say that there is such a demand for clean and healthy beef and poultry in America that you have customers who sign up in advance, come to your farm, and pay you. . . not the market man?" Salatin responded "Yes. That is exactly right." The bureaucrats gazed into the distance and contemplated this apparently novel concept. Another official then pondered out loud *"You mean that the farmer can be the retailer?"*

Ambassador to the United States Serguei Martynov, who assumed his post in 1993, then inquired if Salatin had any special education to prepare him for his vocation. The host responded "No. As a matter of fact, my major in college - a Christian university - was English. There are plenty of colleges that teach conventional agriculture and agribusiness; however, there are no colleges that teach what we do here at Polyface Farm. We produce premium food that commands a premium price. We don't have a problem with salmonella. We don't feed antibiotics. We employ animals in multiple uses and let them perform a lot of the farm labor. We believe in small-scale and high efficiency."

"Where do you get the chicks for your poultry operation?" inquired one official. Salatin responded "We order the day old chicks and they come in the mail." The men looked at each other, then repeated "You have chicks that come in the mail?"

"Exactly" said Salatin. "However, we are implementing a system this year to keep breeder stock on pasture also. We're teaming up with a friend and neighbor to use selective breeding for forage efficiency. This is a new concept in pastured poultry, and we think that we are onto something big with this."

"What about transportation for your poultry to the processing plant?" inquired another delegate.

Salatin responded "We process our chickens right here on the farm. It's clean, it's not overwhelming with the volume we handle, and it's a family effort. Just the five of us can do several hundred in a single morning. Then, in the afternoon, our customers come to the farm, from as far as 200 miles away, to pick up their fresh chicken. Currently we are processing about 8000 birds a year, but we are fortunate in Virginia to be permitted to process up to 20,000 a year without inspection."

Asking about the beef operation that Salatin operates, Chairman Sharetsky inquired about where grain is stored on the farm. "We don't feed our cattle grain. They are grass fed only in an intensive rotational grazing program. Every day, they are turned onto a fresh 'salad bar' of lush new growth. We have used no fertilizer on our pastures for 35 years, and we get 5 times the grass productivity of conventional graziers. This is good for the soil and good for the animal. We have had great success with the effort, and our customers are very happy with the lean and healthy beef we provide." Indeed, Salatin recently released an extensive book on the topic entitled *Salad Bar Beef*. It is the second of his works, the first being entitled *Pastured Poultry Profits*, published in 1993. Ambassador Martynov asked "Who published the book for you?", wherein Salatin replied plainly "I did."

Chairman Sharetsky then turned and spoke directly to his host. His eyes were focused and he spoke with fervor. The assistant to the Chairman of the Supreme Council of the Republic of Belarus then translated "Washington is a very nice capitol because of you. You have to work very hard. I am most impressed with your dedication. The people in Washington owe a lot to Joel Salatin. This country is prosperous because people like you work very hard. In our effort in Belarus to go from collective farming to privatized farms, the people, they are resistant.

"Our people have become too specialized. In the collective farm, one never had to be concerned about buying a tractor. It was provided by the state. And if your job was driving a tractor, then that is all you did and nothing else. The problem confronting us in Belarus is a psychological one. Our people do not know the feeling of being master of your own land. They are unaware of how to deal with farm finances and management. They are too used to being part of a team, and are finding it difficult to be on their own. This free market system is something new, a phenomenon in our country. We must instill in them the desire to learn alternatives, to be able to work on their own, like Joel Salatin."

Chairman Sharetsky summed up the agricultural woes in his country by expressing "Our people. They know *how* to farm, but they do not know *why* to farm."

Appendix C

Resources

Recommended Books:

Backyard Market Gardening: The Entrepreneur's Guide To Selling What You Grow, by Andy Lee

Sell What You Sow! The Grower's Guide To Successful Produce Marketing, by Eric Gibson

Farms of Tomorrow: Community Supported Farms / Farm Supported Communities, by Trauger M. Groh and Steven S.H. McFadden

Salad Bar Beef, by Joel Salatin

Pastured Poultry Profits, by Joel Salatin

The New Organic Grower: A Master's Manual of Tools and Techniques for the Home and Market Gardener, by Eliot Coleman

Backyard Cash Crops: The Sourcebook for Growing and Selling over 200 High-value Specialty Crops, by Craig Wallin

Barry Ballister's Fruit and Vegetable Stand, by Barry Ballister

Market What You Grow: A Practical Manual for Home Gardeners, Market Gardeners and Small Farmers, by Ralph J. Hils, Jr.

Successful Small-Scale Farming: An Organic Approach, by Karl Schwenke

Growing for Market: On a Few Acres or in Your Backyard, edited by Roger B. Yepson, Jr.

Ten Acres Enough: The Small Farm Dream is Possible, by Ralph C. Miller and Lynn R. Miller

How To Win Friends and Influence People, by Dale Carnegie

A Garlic Testament, by Stanley Crawford

The Flower Farmer: An Organic Grower's Guide to Raising and Selling Cut Flowers, by Lynn Byczynski

Marketing Your Produce: Ideas for Small-Scale Farmers, by the publishers of *Growing For Market*

Recommended Magazines:

Acres USA
P.O. Box 8800
Metairie, LA 70011

American Small Farm
9420 Topanga Canyon
Chatsworth, CA 91311-5759

Coming Home
P.O. Box 367
Savannah, TN 38372

Country Journal
P.O. Box 420235
Palm Coast, FL 32142-0235

Countryside and Small Stock Journal
W11564 Hwy 64
Withee, WI 54498

Growing for Market
P.O. Box 3747
Lawrence, Kansas 66046

Organic Gardening
33 E. Minor Street
Emmaus, PA 18098

Quit You Like Men
P.O. Box 1050
Ripley, MS 38663-9430

Small Farmers' Journal
P.O. Box 1627
Sisters, OR 97759

Small Farm Today
3903 W. Ridge Trail Rd.
Clark, MO 65243-9525

Stockman Grass Farmer
5135 Galaxie Drive
Suite 300C
Jackson, MS 39206

Recommended Catalogs & Directories:

1996 National Farmers' Market Directory
USDA/AMS/TMD/W&AM
Room 2642-South
P.O. Box 96456
Washington, D.C. 20090-6456

Brittingham Plant Farms
P.O. Box 2538
Salisbury, MD 21802

DripWorks
380 Maple Street
Willits, CA 95490

Garden's Alive
5100 Schenley Place
Lawrenceburg, In 47025

Harmony Farm Supply & Nursery
3244 Hwy. 116 No. G
Sebastopol, CA 95472

Johnny's Selected Seeds
Foss Hill Road
Albion, ME 04910-9731

Lehman's Hardware
One Lehman Circle, Box 41
Kidron, OH 44636

A.M. Leonard
P.O. Box 816
Pique, OH 45356

Mainline Equipment
P.O. Box 526
London, OH 43140

Marti Poultry Farm
P.O. Box 27
Windsor, MO 65360-0027

Mellinger's
2310 W. South Range Rd..
North Lima, OH 44452-9731

NASCO Farm & Ranch
P.O. Box 901
Fort Atkinson, WI 53538-0901

Peaceful Valley Farm Supply
P.O. Box 2209
Grass Valley, CA 95945

Ridgway Hatcheries
P.O. Box 306
Larue, OH 43332-0306

Rispens Seeds
P.O. Box 5
Lansing, IL 60438

Seeds of Change
P.O. Box 15700
Sante Fe, NM 87506-5700

R. H. Shumway Seedsman
P.O. Box 1
Graniteville, SC 29829

Southern Exposure Seed Exchange
P.O. Box 170
Earlysville, VA 22936

Stokes Seeds
P.O. Box 548
Buffalo, NY 14240

Territorial Seed Company
P.O. Box 157
Cottage Grove, OR 97424

Totally Tomatoes
P.O. Box 308
Bomoseen, VT 05732-0308

Vermont Bean Seed
78 Garden Lane
Fair Haven, Vermont 05743

Appendix D

SAMPLE FARMERS' MARKET RULES AND GUIDELINES

FARMERS' MARKET # 1 (Less than 20 farmers)

1. The farmers' market will operate from 9:00 A.M. until 12:00 noon each Saturday in July, Augusta, and September at the Grange Hall on Main Street.

2. Participation is open to any person interested in its promotion, and who abides by the Market rules.

3. Vendors must arrive at or before 9:00 A.M. for registration and setup. After checking in, vendors will be assigned a selling space on a first-come, first-served basis by the Market Master.

4. Vendors will be charged 10% of the amount they make from sales each market day. Vendors will be responsible for figuring their own total daily sales, and calculating the 10%. Payment must be made on market day. The proceeds of this fee will be used to promote and operate the Farmers' Market.

5. Each vendor will be responsible for setting up, displaying, and packaging his or her products, as well as protecting those products from the elements. Each vendor must leave his or her selling area in clean and orderly condition. All refuse and unsold goods must be removed from the market area by the vendor.

6. Small vendors, such as gardeners, may coordinate and sell for each other.

7. Only locally grown or produced food products, flowers, herbs, and baked goods may be offered for sale. Purchased products may not be sold.

8. Baked and processed goods (such as bread, jellies, jams, preserves, and pickles) must be prepared by the vendor, and the vendor is responsible for any necessary licenses.

9. Prices should be fair market value, negotiated by the vendor and the customer. The Market Master will make a copy of the Weekly Farmers' Bulletin available as a general guide for pricing. The Farmers' Market is not responsible for arrangements made between customers and vendors. No warranty of any sort, expressed or implied, is made by the Farmers' Market.

10. The Farmers' Market Committee will assign a Market Master on a rotating basis to work with vendors. The Market Master is the official representative of the Market Committee. If problems arise on market day, they will be resolved by the Market Master.

11. No goods are to be sold before the market officially opens.

12. Surplus food may be donated to local charities. The Market Master will make every effort to see that needy groups are facilitated on a fair basis.

13. Any accident or injury must be immediately reported to the Market Master. Anyone who comes to participate in the market, vendor or customer, comes at his or her own risk. The Farmers' Market is not liable for an injury to person or property.

14. Willful violation of the market rules may subject a vendor to exclusion from further participation in the market. All violations will be reviewed by the Market Committee.

15. These rules are intended to be fair and in the best interests of all who participate in the Farmers' Market. The Market Committee may, at any time, modify or add to these rules to better serve those interests.

FARMERS' MARKET #2
(Less than 30 farmers)

Foreword

The beginning of the new season for the Farmers' Market is rapidly approaching. As you will note, the season begins on Saturday, April 13, and continues through Tuesday, November 26.

Attached you will find a copy of the new Guidelines and an Application for Reserved Space. Should you have any questions concerning either of the forms, fell free to call the Market Master.

The success of our Market depends on a good growing season and active participation by many and varied vendors. The Board looks forward to working with each of you for the successful marketing of our farm products. Should you desire guidance with regard to the types of products which may have customer appeal, please contact any Board member and we will try to help you in this regard.

Farmers' Market Information and Guidelines

1. Location: Lower East side of the Water Street Parking Deck

2. When: Every Tuesday and Saturday from 7:00 A.M. until 1:00 P.M., starting Saturday, April 13 (weather permitting) and running through Tuesday, November 25 (weather permitting).

3. What May Be Sold: Farm and kitchen products, plants, and handcrafts are suitable items. Products sold must be grown or produced by the licensed producer. Meat, fish, fowl, butter, pickles, cheese pies, custards or other products containing an egg base may not be sold. For more information and requirements regarding products from the home kitchen, you may contact the State Bureau of Food Inspection. Participants must comply with state and local health regulations. The State Department of Agriculture has imposed some restrictions on home canned products such as tomatoes.

4. Permits and Fees: Sellers of farm and kitchen products must obtain a Producer's Permit at no charge from the Commissioner of Revenue. A *copy* of the permit must then be taken to the police station. Sellers of handcrafts must also purchase a Retail Merchants License from the Commissioner of Revenue. All permits must be displayed by the seller. In addition to the above local permits and fees, the Farmers' Market collects a $20.00 membership fee to be paid on or before your first day at the Market. There is also a daily fee charged according to the following chart:

GROSS SALES	FEES
$0 TO 25.00	$1.00
25.01 TO 50.00	2.00
50.01 TO 75.00	3.00
75.01 TO 100.00	4.00
100.01 TO 125.00	5.00
125.01 TO 150.00	6.00
150.01 TO 175.00	7.00
175.01 TO 200.00	8.00
Etc.	

117

The daily fee will be collected daily for the previous market day and applies to all goods transferred at the Market. This includes advance orders and wholesale. Fees are to be placed in envelopes provided by the Market Master. These fees will be collected by the association of market participants and will be used for advertising and other expenses.

5. Parking and Space Assignments: Vehicles should be parked at right angles to the curb. Please contact the Market Master prior to your first day at the market to be assigned a space. Sellers who want to reserve a specific location for their regular use may fill out an application for a reserved space. There may be an unreserved section with spaces available on a first-come, first-served basis. A producer may request the use of more than one space if needed for high volume sales.

6. Sales Tax: State law requires market participants to register with the State Department of Taxation, and also requires participants to collect and report sales tax collected. Applications and information may be obtained from the Department of Taxation Office.

7. Reserved Spaces: The application for, and assignment of, a reserved space means a commitment to the market under the following terms:

♦ The vendor makes a commitment to be present at least once each week during the period of your active participation.

♦ If unable to attend at least once each week, the vendor is responsible for notifying the Market Master prior to the absence.

♦ Loss of reserved space will be based on irregular attendance or two consecutive absences without notifying the Market Master.

♦ All space assignments will be made by the Market Master based on factors such as the date of request, the length of commitment, frequency of attendance, and the needs of the vendor and the needs of the market.

♦ For those needing additional space for high volume sales, an additional space (or two, if available) may be reserved for $20.00 per space.

♦ Requests to relocate to a different space or complaints should be directed to the Market Master. If you are dissatisfied with the handling of your complaint, it will be reviewed by the Market Board at your request.

FARMERS' MARKET # 3
(less than 40 farmers)

1. A producer's certificate must be complete before selling at the market. They are available at no charge from either the Market Master on location at the Market in the Parking Lot the day the vendor wishes to sell, City Hall, the County Farm Bureau, or the County Extension Office. The purpose of this certificate is to ensure that the produce and food products sold are produced by the vendor, his or her family, or employees. The Market Master will retain this certificate and send a copy to the Commissioner of Revenue's Office. This allows the producer to be exempt from purchasing a city business license for the purpose of selling at the Farmers' Market. The certificate is only good for one season.

2. The market will be open for retail sales between the hours of 7:00 A.M. and 12:00 noon on each Saturday of the market season as designated by the City Manager. The new season is from April 13 through October 26. At the beginning of each market day, sellers shall check in with the Market Master. Vendors should be set up at the market by 6:45 A.M. Selling is allowed only during the scheduled market hours. Each seller is required to stop selling at noon and have all produce, baked goods, boxes, containers, etc loaded for removal by 12:30 P.M. The Market Master shall have the authority to change the market hours on special occasions. The market will be held every Saturday during the season - rain or shine.

3. Producers may sell farm and kitchen products including fruit, vegetables, eggs, cut flowers and potted plants. The following items are subject to the Department of Agriculture approval: cider, jams, jellies, relishes, honey, canned goods and baked goods. Any vendor selling these products must contact the Bureau of Food Inspection for inspection and approval.

4. Arts and crafts generally are not permitted for sale at the Market unless they are made from agricultural products and approved by the Market Master and the Market Committee.

5. In general, only producers may sell at the market. Producer is defined as the person that grows or makes the product and may also include the producer's immediate family, partners, or employees or local cooperatives upon review. Any complaints filed with the Market Master about producers not following this rule will be investigated by the Market Master with the County Extension Agent and may result in expulsion from the market for the remainder of the selling season.

6. Vendors may rent a reserved space along the brick sidewalk on the first row of spaces in the west side of the parking lot for the entire season for the cost of $35.00. If a vendor wishes a second space in the first row (subject to availability), the cost is an additional $100.00 for the season. Reserved spaces in the second row will be $25.00 each for the entire season. Reserved spaces for the new season will be based on a first-come, first-served basis with the exception that preference will be given to previous season vendors. Within previous seasons vendors, preference will be given to those who attended the most market days. Any available space in the first row of the market may be rented on a daily basis by a vendor without a reserved space for $5.00. Fees for reserved spaces must be paid in full before March 15 of the new season.

7. If the vendor of a reserved space is not at the market by 7:00 A.M., the space may be assigned to another vendor for that particular

day. The late arriving vendor of the reserved space may be assigned to the unreserved section by the Market Master. Call the Market Master when you are running late. If we know you are coming, we will not reassign your space. Also, call the Market Master in advance if you plan to be absent for the market day. Phone numbers are provided for the two phone booths located in the parking lot. Use these numbers if you call after 6:00 A.M. on Saturday.

8. Vendors without a reserved space may park in the unreserved section, which is on a first-come, first-served basis as assigned by the Market Master each week of the market. Anyone arriving after 7:00 A.M. needs to check in with the Market Master before parking. Late arrivals will be assigned a space where setup will make the least impact on the operation of the market and the safety of the customers.

9. The Market Master may designate the vending location for all vendors with or without vehicles; and if the Market Master judges it necessary, may request a vendor to relocate.

10. Vendors must sell from only one vehicle and must occupy only one space as assigned by the Market Master unless a second space has been assigned and paid for. Maximum vehicle size is a I-ton pickup truck or van without making prior arrangements. All displays, including umbrellas and canopies, must be securely anchored and must not extend beyond the limits of the assigned space. Umbrellas and canopies must not be anchored *into* the asphalt parking lot surface. If the vendor is using a trailer to display and sell goods, the towing vehicle must be disconnected and moved out of the area.

11. Upon checking out, all vendors will pay the Market Master 5 ½% of their daily sales for their slot fee, and 4 ½% of sales for state sales tax unless they are exempt by payment directly to the state. These fees are paid every week the vendor sells at the market and are in addition to any seasonal one-time reserve space fee that a vendor may pay (see rule #6). If the vendor pays directly to the state, the vendor will provide the Market Master with his or her Sales Tax Identification Number.

12. Each vendor in the market area shall be solely responsible at all times for the cleanliness around the vending the area regardless of the origin of the debris in that location. We encourage each vendor to bring a trash container and a broom in order to leave the space clean at the end of the day.

13. It is recommended that all displays and food items be raised at least 24 inches from the ground.

14. Vendors are discouraged from bringing pets. If you choose to bring one, it must be restrained in a way that is not disruptive to the market. Vendors are responsible for any damage that is caused by a pet.

15. No poultry, game, or livestock shall be slaughtered or dressed within the market area.

16. The sale of live animals is not allowed at the market.

17. No cooking within the market area is permitted without prior approval from the Market Master and the Department of Health.

18. The sale of cooked meats, game, poultry or seafood is allowed if the food comes from a Department of Agriculture inspected and approved kitchen and is held at proper temperature at the market.

19. In order to sell raw, fresh, or frozen meats, game, poultry or seafood, each item must be raised by the producer, must be labeled with a USDA sticker describing item and weight, and must be maintained at an approved temperature at the market. Eggs do not require a USDA sticker, but they do need to be held at proper temperature.

20. Commercial enterprises such as area restaurants and bakeries may sell their baked goods if they produced them. Nurseries may sell bedding plants as long as they make or grow the products offered for sale. Other agricultural items that may be processed by commercial enterprises (wineries, flour mills, coffee roasters, and peanut roasters, etc) may be sold at the market only after review and approval by the Market Committee.

21. Wild flowers or plants sold at the Farmers' Market must not be on the state's list of rare wild plants, which is published by the Division of Natural Heritage. See the Market Master for a current list.

22. Potted plants and flowers for sale must be grown by the seller. No plants may be sold that were removed from public parks or interstate highways.

23. Public rest rooms are located across the street from the market area in the City Parking Garage. There is water on the site.

24. Generators are not allowed at the market. There is a limited number of power outlets available for $3.00 per week (payable to the City via the Market Master). These are on a first-come, first-served basis. Interested vendors must bring their own all-weather extension cords and must park in a specially designated area near the power outlet.

25. Solicitation for products, services, or charitable contributions not specifically addressed as a market commodity or by vendors other than market vendors is not permitted without prior permission of the City Manager.

26. Children under the age of fourteen years shall not be a vendor unless accompanied by an adult responsible for the child's conduct and safety.

27. Fraudulent, dishonest, or deceptive merchandising or collusion to set prices among vendors may be grounds for forfeiture of the right to do business of any kind in the market for a length of time to be determined by the Market Master and the Market Committee.

28. It is the responsibility of each producer to abide by all state and federal regulations which govern the production, harvest, preparation, preservation, labeling, or safety of the product the vendor offers for sale at the market.

29. Up to two non-profit organizations per week may sell at the market (one bake sale and one selling breakfast food). Each non-profit organization must pay a flat $5.00 fee per week, and an additional $3.00 per week if they use electricity. Reservations for non-profit organizations are honored on a first-come, first-served basis.

30. At the annual organizational meeting held prior to the opening of the market season, yearly elections are held for the two vendor representatives to the Committee. The representatives for this season are _____ and _____.

31. The Market Master shall have the full power to enforce all rules and regulations within the market area as authorized by the appropriate section of the City Code. Failure by

any vendor to comply with any of these rules and regulations can result in the forfeiture of the right to do business of any kind in the market for a length of time determined by the Market Master and the Market Committee.

Index

126

Reader Feedback

Just like you ask your farmers' market customers for feedback (you *do*, don't you?), I would appreciate *your* feedback to help update and improve this book in the future. Please take a moment of your time and put your thoughts in writing. If you run out of room, please feel free to use additional paper. Your comments are welcome.

Things I like about *Dynamic Farmers' Marketing* include:

_____

_____

_____

_____

_____

_____

_____

_____

_____

_____

_____

Dynamic Farmers' Marketing could be improved by:

_____

_____

_____

_____

_____

_____

_____

_____

_____

Send comments to:
Bittersweet Farmstead, P.O. Box 52, Middlebrook, VA 24459

128

Purchasing Information

Additional copies of *Dynamic Farmers' Marketing: A Guide to Successfully Selling Your Farmers' Market Products* may be purchased directly from:

Bittersweet Farmstead
P.O. Box 52
Middlebrook, VA 24459

(540) 886-8477

PRICES

All prices are in U.S. funds. For international orders, contact the author.

# of Books   ~~ling~~
1
2-5
6 or more

For further info about bulk orders,
see our web site:
www.emarketfarm.com
or call (540) 248-3938

Additional shipping: For six or more, add $.75 s/h for each copy *over* six.

Example: Place a bulk order for your farmers' market and purchase 10 copies.

        10 copies   x $11.95   = $119.50
        Plus reg s/h   7.00
        Plus additional s/h (4 x .75) ___3.00
                                            $129.50

Note: Virginia residents must include 4 ½ % sales tax for books (not shipping).

Inflation is a fact of life and costs of printing and production may increase; therefore, the author reserves the right to alter pricing structure at any time.